Making Plant Life Easier and Productive Under Salinity - Updates and Prospects

Edited by Naser A. Anjum, Asim Masood, Palaniswamy Thangavel and Nafees A. Khan

Published in London, United Kingdom

Making Plant Life Easier and Productive Under Salinity - Updates and Prospects
http://dx.doi.org/10.5772/intechopen.106124
Edited by Naser A. Anjum, Asim Masood, Palaniswamy Thangavel and Nafees A. Khan

Contributors
Rania Djemal, Moez Hanin, Chantal Ebel, Maneesh Kumar, Himanshu Bharadwaj, Komal Kumari, Ram Baran Singh, Rajni Devi, Emine Ayaz Tilkat, Ayşe Hoşer, Veysel Süzerer, Engin Tilkat, Chandima Kumudini Ariyarathna Hanchaplola Appuhamilage, Bhagya Indeevari Dasanayaka, Demuni Sumith de Zoysa Abeysiriwardena, Naser A. Anjum, Asim Masood, Palaniswamy Thangavel, Nafees A. Khan, Faisal Rasheed

First published in London, United Kingdom, 2023 by IntechOpen
IntechOpen is the global imprint of INTECHOPEN LIMITED, registered in England and Wales, registration number: 11086078, 5 Princes Gate Court, London, SW7 2QJ, United Kingdom

British Library Cataloguing-in-Publication Data
A catalogue record for this book is available from the British Library

Additional hard and PDF copies can be obtained from orders@intechopen.com

Making Plant Life Easier and Productive Under Salinity - Updates and Prospects
Edited by Naser A. Anjum, Asim Masood, Palaniswamy Thangavel and Nafees A. Khan
p. cm.
Print ISBN 978-1-83768-876-0
Online ISBN 978-1-83768-877-7
eBook (PDF) ISBN 978-1-83768-878-4

For EU product safety concerns:
IN TECH d.o.o., Prolaz Marije Krucifikse Kozulić 3, 51000 Rijeka, Croatia,
info@intechopen.com or visit our website at intechopen.com.

Meet the editors

Dr. Naser A. Anjum obtained a Ph.D. in Botany from Jamia Hamdard, New Delhi, India, and received post-doctoral research training at the University of Aveiro, Portugal; the Agricultural Biotechnology Research Centre, Academia Sinica, Taiwan; and Aligarh Muslim University, Aligarh, India. Dr. Anjum has made significant contributions to plant/crop environment adaptation biology. He has received prestigious research awards from reputed funding agencies including the Portuguese Foundation for Science and Technology (FCT), the Council of Scientific & Industrial Research (CSIR), and the Department of Biotechnology (DBT), Government of India. Dr. Anjum has published over 125 peer-reviewed papers and edited 18 books and journal special issues. He is an elected fellow of The Linnean Society, London; Indian Botanical Society; International Society of Environmental Botanists, India; and National Academy of Biological Sciences, India.

Dr. Asim Masood has unveiled major insights into abiotic stress (metals, salt, and temperature) impacts and underlying physiological/biochemical mechanisms in crop plants. He received a Start-Up Career Grant from the University Grants Commission, New Delhi, India, and the DST-SERB Young Scientist Project from the Department of Science and Technology, New Delhi, India. His major research contributions include the involvement of ethylene in sulfur-mediated alleviation of cadmium, ethylene-mediated regulation of photosynthesis as a result of reduced glucose sensitivity, ethylene-mediated modulation of gibberellic acid-induced sulfur-assimilation under cadmium stress, and ethylene-mediated reversal of salinity-impacts involving ethylene-abscisic acid coordination. Dr. Masood has published more than sixty research papers.

Dr. Palaniswamy Thangavel specializes in soil ecology and phytoremediation, biofortification, and biofuels. He has contributed significantly to the environmental sciences, where his works have unveiled insights into the mechanisms of toxic metals/metalloids accumulation and tolerance in economically important crops, the phytoremedial potential of several metal-tolerant flora, phytochelatins and polyamines in metal detoxification mechanisms in various tree/hyperaccumulator species. With more than 15 years of research experience, Dr. Thangavel has handled six research grants, published more than fifty refereed papers, and edited several reference books and monographs. He has also been a member of several professional societies including the American Society of Plant Biologists and the International Phytotechnology Society.

Prof. Nafees A. Khan has contributed to plant abiotic tolerance mechanisms vis-à-vis a positive shift in the source-sink relationship. A modulation in the ascorbate-glutathione system was suggested in plants as a result of the induction of signal transduction by ethylene and its coordination with other major phytohormones. His groundbreaking R&D contributions in physio-biochemistry of abiotic stress tolerance strategies earned publications in high-impact factor journals. Prof. Khan was declared a Highly Cited Researcher by the Web of Science consecutively four times from 2019 to 2022 and awarded the 2023 Clarivate India Research Excellence Citation Award. He has edited twenty-one books and served as editor/guest editor of leading publishing platforms. Prof. Khan is elected Fellow of the National Academy of Sciences, India (NASI), The Linnean Society, the Indian Society for Plant Physiology, and the Indian Botanical Society.

Contents

Preface

Among the major devastating abiotic challenges is the salinization of soils, which are natural finite resources sustaining most life on Earth. Soils exhibiting saturation paste extracts (ECe) in the root zone as electrical conductivity (EC) > 4.0 dS m−1 (≈ 40 mM NaCl; dSm-1 = decisiemens per meter) at 25°C and 15% exchangeable Na+ ion are considered saline. By the year 2050, a projected increase of 16.2 million ha will be witnessed in salt-affected areas. In terms of tolerance to salinity levels, plants are grouped into two categories, namely halophytes (salinity tolerant and adapted to salinized environments) and glycophytes (salinity-sensitive plants). Unfortunately, most agricultural crop plants are glycophytes (salt sensitive), where soil salinization adversely inhibits the plants' absorption of water and nutrients and eventually severely impairs the major physiological/biochemical and molecular attributes of crop plants as well as their yield (30%–50% reduction). Thus, the salinity sensitivity of most crop plants, increasing rate of land salinization, and salinity-caused losses in crop productivity are challenging food security.

Interestingly, the life of plants under salinity can be made easier and more productive by employing strategies comprising optimum and timely supply of mineral nutrients, compatible solutes, bio-stimulants, nanomaterials, phytohormones, phenolic compounds, and microorganisms, as well as considering the crosstalk of mineral nutrients/biostimulants–phytohormones and molecular-genetic approaches. These approaches, in isolation and/or combination, may boost the efficiency of the plant's inherent mechanisms for its improved sustenance and productivity under salinity stress.

Making Plant Life Easier and Productive Under Salinity – Updates and Prospects introduces the concept of salinity, its major impacts, and important approaches for making plant life easier and more productive under salinity; reviews the scale and complexity of salinity impacts on Sri Lankan rice farming systems, along with presenting major actionable insights; reviews the salinity stress responses of major metabolites and in vitro production of terpene in plants; discusses the major mechanisms underlying phytohormone-mediated control of salinity impacts in plants; and appraises the literature on genomics-assisted breeding approaches for achieving salinity tolerance in cereal crops.

We believe that this volume will serve as an important resource for plant biologists and agriculturists as well as research students.

We would like to thank our publisher IntechOpen for providing a platform for showcasing our work. We are also thankful to Publishing Process Manager Mrs. Karla

Skuliber for her efficiency and guidance, which helped us to accomplish this book project. Our sincere gratitude also goes to the chapter authors who contributed their time and expertise. Without their support, this book would not have become a reality.

Naser A. Anjum, Asim Masood and Nafees A. Khan
Department of Botany,
Aligarh Muslim University,
Aligarh, India

Palaniswamy Thangavel
Department of Environmental Science,
Periyar University,
Salem, India

Chapter 1

Introductory Chapter: Making Plant Life Easier and Productive under Salinity – Updates and Prospects

Naser A. Anjum, Asim Masood, Palaniswamy Thangavel and Nafees A. Khan

1. Introduction

1.1 Human, food and plants

Almost all life forms on the Earth fundamentally depend on plants and/or plant-based (food and non-food) resources. Human being, in particular, are indebted (directly or indirectly) to the diversity of plants for oxygen, food, fuel, fiber, medicines, and even for shelter. Being sessile by nature, plants have to closely interact with the immediate environment (growing conditions comprising air, water, soil, energy/light) to sustain their own lives and accomplish the aforementioned notable contributions to human life. As an enigmatic, heterogeneous, multiphasic and porous system, soil acts as a natural growth environment/medium for diverse land plants. The health of soil is closely linked with the health of plants and that of their immediate and long-term consumers. Interestingly, 80% of average calorie consumption comes from crop plants grown directly in soil. Hence, healthy soil has been widely argued vital both as a resource for feeding the burgeoning global population via agriculture, and also for realizing most of the United Nations Sustainable Development Goals [1–5].

1.2 Soil salinization: concept and types

The health of soil is being significantly impacted by increased salinization due to excess accumulation of varied salts (e.g., cations: Na^+, Ca^{2+} and Mg^{2+}; anions: Cl^-, SO_4^{2-}, CO_3^{2-} and HCO_3^-). Based mainly on the soil properties, namely electrical conductivity (EC), pH, exchangeable sodium (Na^+) percent (ESP), Na^+ adsorption ratio (SAR), total soluble salts (TSS), and total dissolved solids (TDS), soils can be categorized in three major types: saline, sodic and saline-sodic. EC of the saturation paste extract (EC_e) is the measure of salinity, whereas the measure of sodicity is the exchangeable sodium (Na^+) percentage (ESP) or the sodium adsorption ratio (SAR). Soils can be saline (ECe > 4 dS m^{-1} (decisiemens per meter)) at 25°C and ESP < 15 (high soluble salts and low exchangeable Na^+; pH 7.0–8.5), sodic (Ece < 4 dS m^{-1} and ESP > 15; with a high amount of exchangeable Na^+ ions on the cation-exchange sites; weak bond between soil particles; pH > 8.5) or saline-sodic (Ece > 4 dS m^{-1} and ESP > 15; both salts and exchangeable sodium are high). According to the first ever country-driven global map of salt-affected soils (GSASmap, V1.0.0), which comprised over 118 countries with 257 and 419 locations (covering 85% of global land area), 85% of salt-affected top-soils are saline, 10% are sodic and 5% are saline-sodic, 62% of salt-affected subsoils are saline, 24% are sodic and 14% are saline-sodic (FAO-GloSIS 2023 [6]; **Figure 1**). Major consequences of sodicity and salinity on soil health are summarized in **Figure 2**.

The building-up of elevated level of varied salts (e.g., cations: Na^+, Ca^{2+} and Mg^{2+}; anions: Cl^-, SO_4^{2-}, CO_3^{2-} and HCO_3^-) in the soils may be caused naturally (*leading to primary salinity*) or human-induced (or anthropogenic activities) (*leading to secondary salinity*). Interestingly, the natural climate conditions; different geological, hydrological and pedological processes; wind; rainfall; parent rock weathering; long-term natural accumulation of salts (including Cl^- of Na^+, Ca^{2+} and Mg^{2+} and sometimes SO_4^{2-} and CO_3^{2-}); and higher evapotranspiration (versus precipitation) largely contribute to the primary or natural salinity in the soil or surface water. On the other hand, the human-induced (or anthropogenic) activities done in agricultural management practices are largely inappropriate, which cause poor drainage and arbitrary irrigation, disrupt the hydrologic balance of the soil between water applied (irrigation or rainfall) and water used by crops (transpiration), leading ultimately to the secondary salinization, the major cause of the loss of agricultural soils (**Figure 3**) [6, 8]. Both primary and secondary

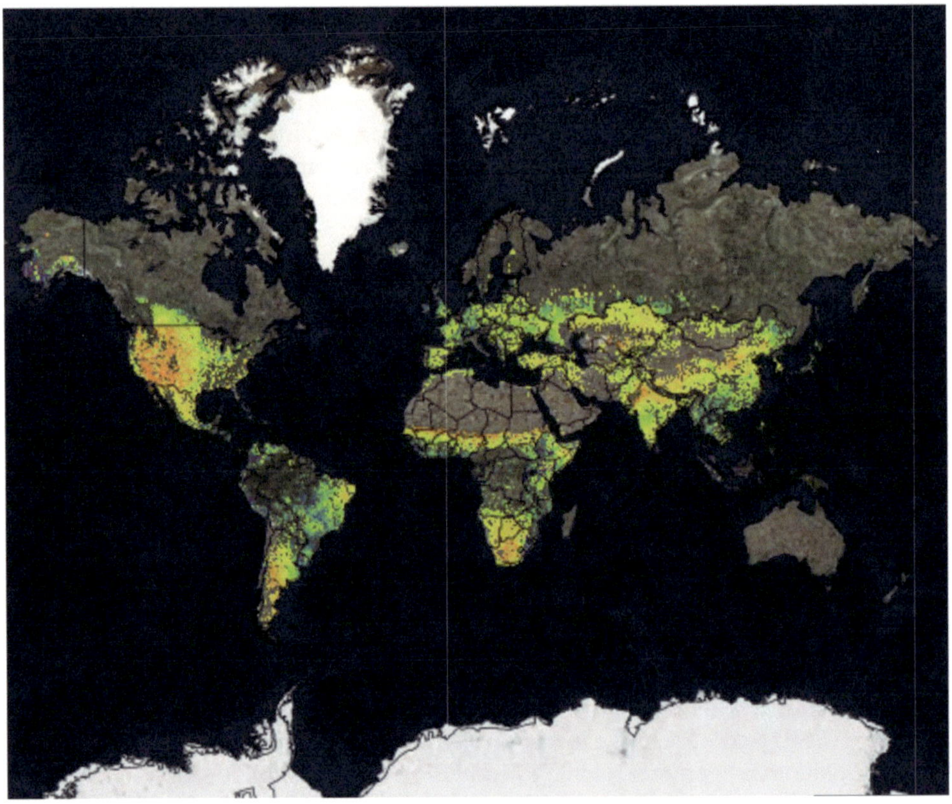

Figure 1.
Global map of salt-affected soils (GSASmap) [6].

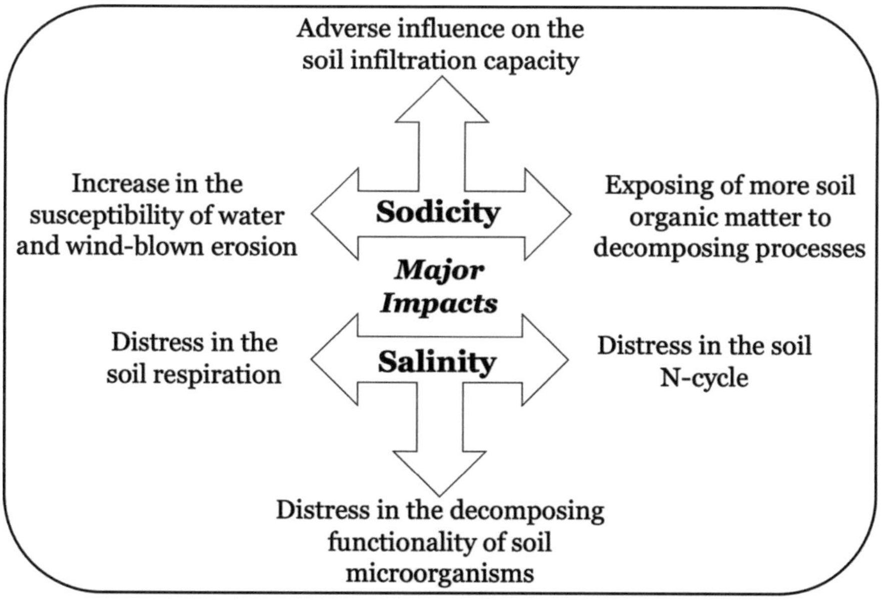

Figure 2.
Schematic representation of the major impacts of sodicity and salinity on soil health [7].

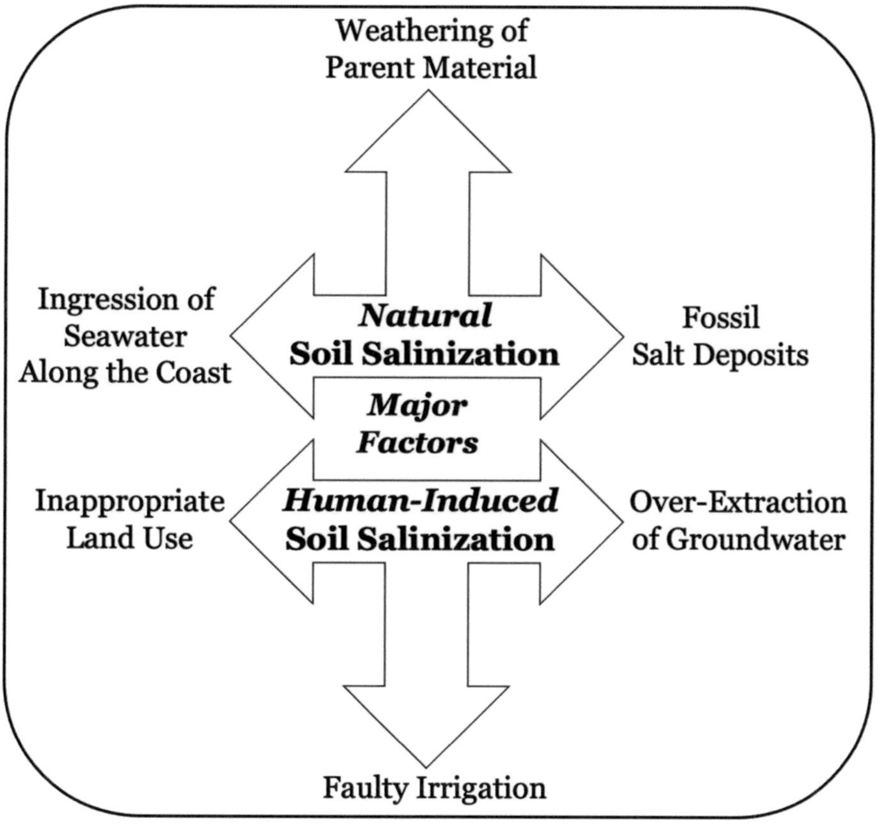

Figure 3.
Schematic representation of the major factor contributing to natural and human-induced soil salinization [7].

salinization of soils mainly occurs in arid and semi-arid regions, where precipitation to evapotranspiration ratio is low. Notably, climate change is among the major factors known to influence the global distribution of salt-affected soils [9–13].

Notably, soil salinization is at the top among the major climate change-influenced abiotic stress factors and is also known to bring severe consequences at both agricultural soil and agricultural crop plant levels. It significantly changes the major physiochemical and biological characteristics (including soil structure, soil microbial activity, etc.). These soil-level changes culminate in the inhibition of absorption of water and nutrients by plants and severely impairing the physiological/biochemical, molecular and yield (≈50% reduction) attributes of crop plants. On the other hand, agriculture, in particular, is under pressure to accelerate crop plant yield (by >70%) in order to feed the burgeoning world population that is projected to stabilize at around 9.7 billion by the year 2050 [14].

Given above, efforts must be made to make crop plant life easier and more productive under rapidly increasing soil salinity. It is required to consider a multi-level approach comprising monitoring, assessment, and the management of soil salinization; getting insights into the crop plant physiological/biochemical and molecular-genetic responses to soil salinity; and dissecting the strategies for strengthening plant/crop salinity-tolerance mechanisms.

2. Salinity-impacts in plants

Three major physiological stresses encountered by salinity-exposed plants include ion (Na^+ and Cl^-) toxicity; physiological drought due to low osmotic potential; and nutrient imbalance in plants (**Figure 4**) [15–17]. Salinity-accrued decrease in plant growth and photosynthesis and related variables may also be possible as a result of hyperionic and hyperosmotic stress [18]. Plant tolerance to salinity of the soils of the root zone greatly varies with plant types and their age, and the salinity concentration and exposure duration. However, exhibition of yield reduction was noted in most crop plants at the electrical conductivity (EC) of the saturation extract (Ece) in the root zone >4.0 dS m^{-1} (≈40 mM NaCl; exchangeable sodium of 15%) at 25°C [19–21]. Taking into account representative studies, the major impacts of salinity on plant growth, photosynthesis, yield, nutrient uptake and metabolism, water status, oxidative stress, and antioxidant metabolism are briefly overviewed hereunder.

2.1 Growth, photosynthesis and yield

Soil salinization impacts almost every stage of plant growth and development. Salinity impact on germination has been extensively reported in several test plants [22, 23]. Contingent to salinity concentration, type of salt present, type of plant species, age of test plant, and time of exposure were reported to control the extent of salinity impacts. The major growth traits, namely the rate, percentage and index of germination; leaf area; length, fresh and dry weights of root and shoot; and plant dry mass, were varyingly decreased in salinity-exposed plants [23–25]. Salt stress impacts on photosynthesis and involved several mechanisms. Salinity-accrued impediment in photosynthesis is caused by decreased leaf area and stomatal conductance, declined CO_2 availability and assimilation; Cl^-led inactivity of RuBisCO; degraded D1 and D2 proteins of PSII reaction center; diminished activity of enzymes involved in photosynthetic pigment synthesis; low uptake of Mg^{2+}, and destruction of pigment-protein

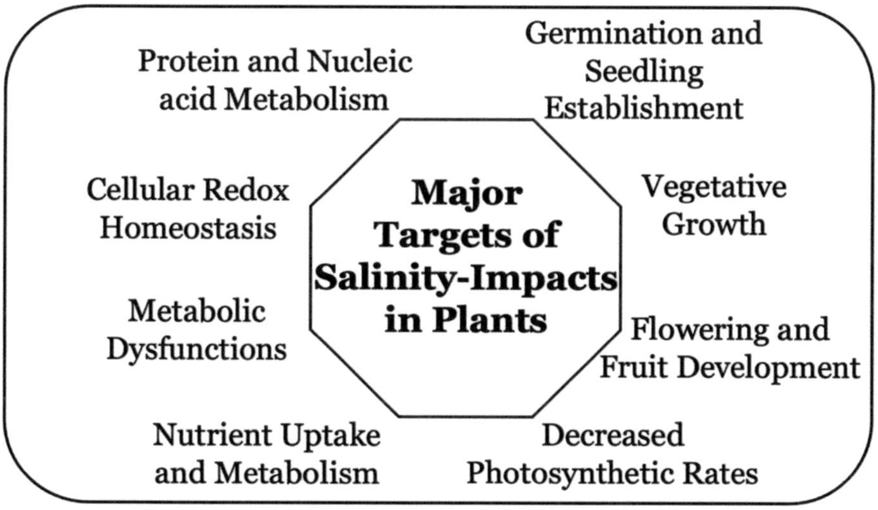

Figure 4.
Schematic representation of the major targets of salinity impacts in plants.

complexes [26, 27]. Elevation in the soluble salts in the rhizosphere is bound to result in brutal yield losses in most crops, which was widely argued to involve salinity impact on different yield components [28, 29].

2.2 Nutrients uptake and metabolism, and water status

Elevated soil salinity significantly impacts the acquisition of most mineral nutrients including Ca, Cu, Fe, N, P, K, S and Zn. Decreased solubility and mobility of Cu and Fe were earlier noted in plants under salinity stress [30, 31]. In fact, salinity conditions tend to immobilize nitrate (NO_3^{2-}; by Cl^-) and ammonium (NH_4^+; by Na^+) ions, the major plant-absorbed N forms [32]. Unavailability of soil-P to plants, its deficiency therein are caused by salinity [33]. Impaired uptake and metabolism of Ca, K, P, N and S were observed in salinity-treated plants, which mainly involved salinity-accrued changes in soil solution's osmotic potential and/or the activity of K^+-selective ion channels [22, 34]. Salt-affected plants exhibit significantly decreased uptake and use efficiency of B, K and P due mainly to the negative interactions with higher concentrations of cations and anions. Salinity-exposed plants also exhibited inhibited K^+ uptake and decreased K^+/Na^+ ratio, which were argued to involve salinity-mediated depolarization of the plasma membrane potential, activation of voltage-gated guard cell outward rectifying K^+ channels and eventual K^+ efflux [35, 36]. Salinity-caused reduced uptake and deficiency of Mn, and decreased solubility and P uptake were also observed in plants under salinity exposure [37, 38]. In several instances, elevated salinity also impacted the assimilation of both N and S via impacting the major N-S assimilatory enzymes [39–41]. Notably, high salinity was reported to affect the major genes involved in the uptake and transport of N, and assimilation can also be affected by high salinity [42]. Salt stress can also impact nitrification and ammonification [43]. The proportion of N-transport amino acids (such as asparagine, glutamate, aspartate and glutamine) decreased in several salinity-exposed test plants [44, 45]. Notably, physiological drought in plants has been reported due to the imposition of high salt concentration in the rhizosphere, and eventual salt-accrued immobilization of water and its unavailability to the plants [46].

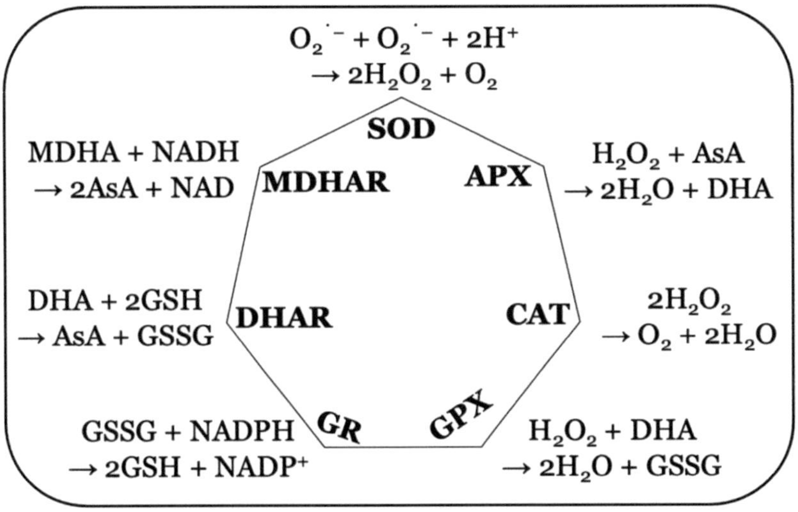

Figure 5.
Schematic representation of the main reactions performed by selected enzymatic antioxidants involved in the scavenging of varied reactive oxygen species (ROS). Abbreviations: DHA, reduced dehydroascorbate; GSSG, oxidized glutathione/GSH disulfide; MDHA, monodehydroascorbate; NADH, nicotinamide adenine dinucleotide; NAD, nicotinamide adenine dinucleotide; NADPH, nicotinamide adenine dinucleotide phosphate; NADP$^+$, oxidized form of nicotinamide adenine dinucleotide phosphate [49–52].

2.3 Oxidative stress and antioxidant metabolism

Elevated soil salinity is widely known to accelerate the accumulation of reactive oxygen species (ROS; such as O_2^-, H_2O_2, •OH, and $^1O_2$) leading to oxidative stress, a physiological condition of imbalance between generation of ROS and their scavenging [22, 47, 48]. Elevated or non-metabolized ROS cause lipid peroxidation and damage macromolecules including DNA and protein. Interestingly, plants possess inherent capacity to activate ROS-scavenging system in order to counteract potential ROS-accrued consequences. Interestingly, plant antioxidant defense system is comprised of enzymatic (superoxide dismutase, SOD; peroxidase, POD; catalase, CAT; ascorbate peroxidase, APX; glutathione peroxidase, GPX; glutathione reductase, GR; monodehydro ascorbate reductase, MDHAR; dehydroascorbate reductase, DHAR); and non-enzymatic (reduced ascorbate, AsA; reduced glutathione, GSH; phenolic, vitamin E, carotenoids and mannitol etc.) components (**Figure 5**) [49–51].

Extensive reports are available on the salinity-mediated elevation in the levels of varied (such as O_2^-, H_2O_2, •OH, and $^1O_2$), and their impact on membrane lipids and leakage of electrolytes have been reported. Salinity stress was also found to induce different components of antioxidant defense system in salinity-exposed plants. Significant enhancements in the activity of enzymes involved in the dismutation of O_2^- (SOD); H_2O_2-metabolism (CAT; APX) and GSH-regeneration (GR) [22, 47, 48].

3. Salinity impacts-minimization in plants

The life of plants under salinity can be made easier considering both plant- and soil-focused strategies, namely adopting approaches aimed at improving plant growth, metabolism, and productivity and employing the management approaches for managing the health (physico-chemical and biological traits) of saline soils.

3.1 Plant health-improving approaches

Plants are endowed with inherent mechanisms for salt stress-impact mitigation, which include hormonal stimulation, ion exchange, antioxidant enzymes and non-enzymes (metabolites) and activation of signaling cascades. However, efficiency of most of these inherent mechanisms can be further improved by employing approaches comprising the optimum and timely supply of mineral nutrients, compatible solutes, bio-stimulants, nanomaterials, phytohormones, phenolic compounds, microorganisms and considering mineral nutrients-phytohormones crosstalk and molecular-genetic approaches.

3.1.1 Mineral nutrients

The role of different mineral nutrients (such as N, P, K and S) in both minimization of salinity impacts and strengthening plant salinity tolerance has been widely studied. The major mechanism involved in the improved health of salinity-treated and N-supplied plants included diminished accumulation of Na^+, efficient N-uptake and assimilation processes, controlled K^+/Na^+ homeostasis, and improved plant-K status [53–55]. S-supply improved plant health and salinity tolerance via maintaining the improved status of S and S-containing compounds (including cysteine, Cys; glutathione, GSH); improved cellular redox homeostasis; efficient ROS-metabolism; decreased oxidative stress; decreased Na^+/K^+ ratio and Na^+ accumulation, increased K^+ and Ca^{2+}; improved uptake of K and P; and reduced electrolyte leakage [22, 48, 56, 57].

3.1.2 Phytohormones

Sustainable improvement of plant growth, metabolism, photosynthesis and productivity (yield) under salinity-affected soils can be possible with the judicious and timely use of various phytohormones and signaling molecules, and thereby minimizing increasing strain on the global food security. The major role (and underlying basic mechanisms) of phytohormones (namely abscisic acid, ABA: a sesquiterpenoid, 15-C compound; auxins: endogenous plant growth regulators; brassinosteroids, BRs: polyhydroxy steroidal phytohormones; cytokinins, CKs: derivatives of adenine or that of phenylurea; ethylene: an unsaturated hydrocarbon gas; and gaseous hormone; gibberellins, gibberellic acid (GA): a large family of tetracyclic di-terpenoid compounds; jasmonic acid, JA: a cyclopentane fatty acid; nitric oxide, NO: a highly versatile gaseous, free-radical, redox-signaling molecule; salicylic acid, SA: a phenolic plant hormone; and strigolactones, SLs: carotenoid derived phytohormone) in improving plant salinity tolerance are briefly highlighted hereunder.

Under saline condition, ABA supply improved plant health by modulating ABA signaling components; reducing Na^+ content, increasing K^+, Mg^{2+} and Ca^{2+} content; improving coordination among antioxidant defense system components; improving cellular level of AsA and GSH; significant reduction in Na^+ content, increasing the contents of hormones such as 1-aminocclopropane carboxylic acid, trans-zeatin, N6-isopentyladenosine, indole-3-acetic acid (IAA); reduction of transpiration flow, regulation of Na^+ ion homeostasis; involving calmodulin signaling cascade; and induction of osmolytes accumulation [58–60]. The supply of auxins has been widely reported to improve plant salinity tolerance involving various mechanisms [61–63]. BRs-mediated improvements in plant salinity tolerance and growth and development mainly involved BRs-supply-induced activity of ROS-metabolizing enzymatic antioxidants (including

APX and CAT), and the cellular levels of non-enzymatic antioxidants (such as AsA and GSH); and decreased electrolyte leakage and membrane lipid peroxidation [24, 64, 65]. Cytokinins-supply can improve plant salinity tolerance via modulating shoot Cl^- exclusion and enhancing antioxidant system and photosynthetic efficiency [66, 67].

Ethylene-induced plant salinity tolerance was argued earlier as a result of ethylene-mediated maintenance of the homeostasis of ions; up-regulation of antioxidant enzymes; improved endogenous ethylene-overproduction; increased activity of seed α-amylase; decreased H_2O_2 and lipid peroxidation; decrease in osmotic stress; PSI cyclic electron flow-mediated controlled non-photochemical quenching [68–70]. Gibberellins (GA)-mediated improvement in plant health involved decreased ion leakage; improved osmolyte accumulation and proline content; elevated Ca^{2+} and K^+ concentrations, and transpiration rates; modulated antioxidants and secondary metabolites; and improved redox homeostasis and coordination among antioxidant enzymes [71–73]. Jasmonic acid (JA)-supply can mediate the salinity-impact mitigation in plants involving increased endogenous levels of cytokinins and auxins, increased α-tocopherol, phenolics, and flavonoids levels; enhanced activity of SOD and APX; increased K^+ and Ca^{2+}; declined Na^+ content; and crosstalk on JA and ABA [74, 75]. Nitric oxide (NO) is the smallest diatomic gas and a gaseous signaling molecule in plants [76, 77]. NO-supply protected plants against salinity impacts by enhancing mineral absorption, maintaining hormone equilibrium, improving osmolyte accumulation, strengthening antioxidative defense systems, mitigating H^+-ATPase inhibition, and maintaining oxidative homeostasis in plants under salt stress [78, 79]. Alleviation of salinity stress in plants has also been achieved through crosstalk of NO with other signaling compounds and phytohormone signaling pathways [80, 81]. In salinity-exposed plants, salicylic acid (SA)-supply reduced Na^+ and Cl^- ions; maintained high GSH level; improved cellular redox environment; regulated AsA-GSH cycle, elevated osmolyte accumulation; and involved characteristic changes in the expression pattern of major GST-gene family members [82–84]. Strigolactones (SLs)-mediated improved salinity tolerance in plants was argued to involve enhanced antioxidant enzyme activity; improved ROS-metabolism, decreased lipid peroxidation and cellular damage; and SLs-ABA-arbuscular mycorrhizal fungi crosstalk [85–88].

3.1.3 Mineral nutrients-phytohormones crosstalk

The major outcomes of the crosstalk of most mineral nutrients with phytohormones have improved plant salinity tolerance. To this end, in several instances, the crosstalk between the major phytohormones (such as indole-3-acetic acid (IAA), gibberellic acid (GA), kinetin (CK), ethylene, 24-epibrassinolide, SA, ABA, brassinosteroid (BR)), with N and S resulted in differential decreasing the content of Na^+ and Cl^- ions; modulation of key enzymes of N and S metabolism; cellular homeostasis, photosynthesis; lowering oxidative damage, improving photosynthetic efficiency, assimilation of N and S, proline content and antioxidant defense system and decreased generation of oxidative stress markers; maintain osmotic balance via controlling cellular osmolytes; modulating the photosynthetic N-use-efficiency and antioxidant metabolism; increased levels of free amino acids and soluble proteins [22, 44, 60, 68].

3.1.4 Osmolytes

Exhibition of almost ceased water flow from soil into roots and cellular dehydration as a result of decreased cellular turgor pressure are common in plants under salinity stress. To counter this, plants employ the mechanism of osmoregulation, which involves the accumulation of varied small water-soluble organic non-toxic solutes (osmolytes/compatible solutes). Osmolytes/compatible solutes such as polyamines (PAs), glycine betaine, b-alanine betaine, dimethyl-sulfonio propionate, and choline-O-sulfate are 'ammonium compounds'; whereas fructan, trehalose, mannitol, D-ononitol and sorbitol are grouped as the 'sugars and sugar alcohols'. On the other, the list of major amino acids acting as osmolytes includes proline and ectoine. The accumulation of most of these solutes in the cytoplasm largely leads to the maintenance of low cytosolic and cytosolic concentrations within narrow limits (100–150 mM) across a broad range of external and vacuolar concentrations of NaCl [89–91]. In addition to performing cytoplasmic osmoregulation/osmotic adjustment, osmolytes have been considered as efficient oxidative stress-busters in plants under varied abiotic stresses (**Figure 6**) [92].

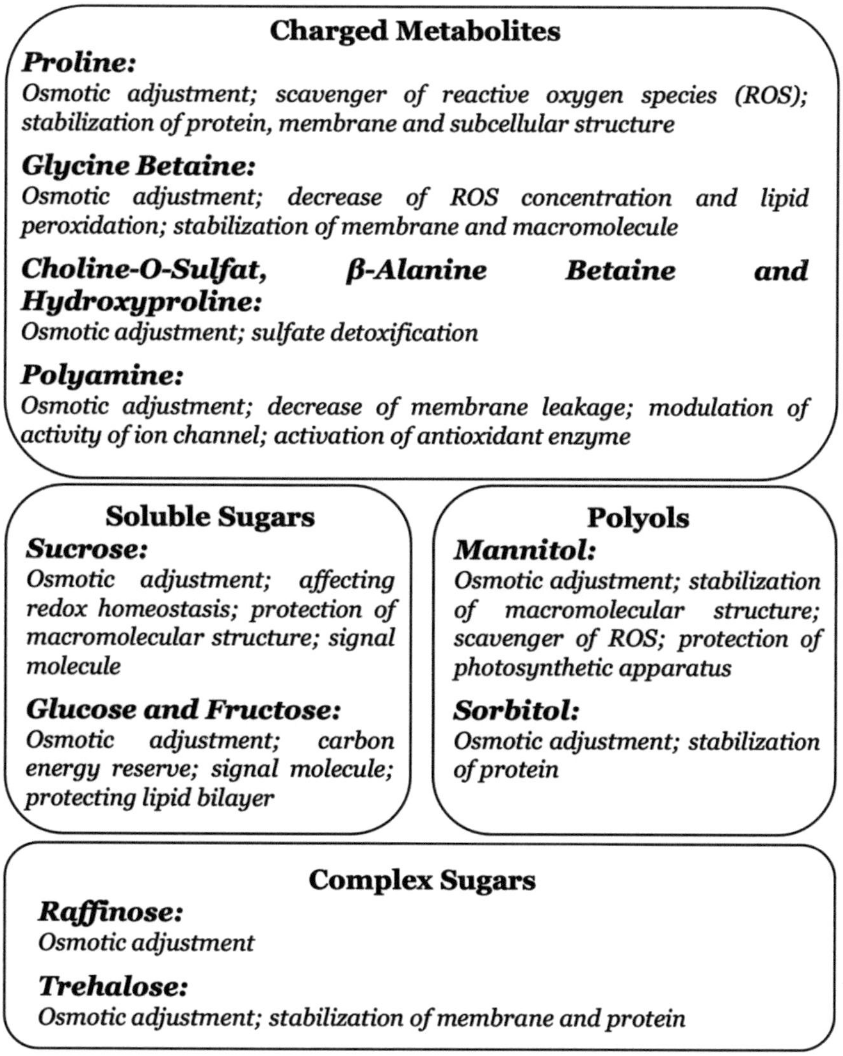

Figure 6.
Schematic representation of the major osmolytes involved in osmotic adjustment in plants under salinity stress [52].

Salinity-exposed plants have been extensively reported to accumulate polyamines (PAs) which are low-molecular-weight aliphatic amines/polycations [91, 93, 94]. Notably, the list of most common natural PAs present in plants includes Spermidine (Spd), spermine (Spm) and putrescine (Put). These PAs are known to act as signaling molecules in plant-environmental stresses [95]. PAs-supply mediated strengthening of antioxidant defense system has been reported in plants [36, 96]. The synergistic effect of PAs (Put + Spd) was reported to confer salinity tolerance in foxtail millet (*Setaria italica*), by inducing antioxidant enzymes and osmoprotectants, and maintaining coordination among the complex physiological and biochemical processes [97]. Elevation in the cellular level of proline is involved plant salt stress tolerance mechanisms. Beside delivering C, N, and energy during stress condition, proline scavenges varied ROS and also stabilizes DNA, proteins and membranes, and reduces NaCl-induced enzyme denaturation [91, 98]. Osmotic adjustment in salinity-impacted plants can also be accomplished with the accumulation of total soluble sugars, namely glucose, sucrose, dextrins, and maltose, where these sugars provide osmoprotection and also act as C-storage [99]. A non-reducing storage disaccharide, trehalose is mainly involved in the regulation of carbohydrate metabolism [100]. Interestingly, trehalose-mediated improvement in plant salinity tolerance involves the maintenance of K^+:Na^+ ratio, ROS-scavenging ROS, and increased soluble sugar concentration [101–103]. Mainly acting as important osmolytes in plant vacuoles, organic acids can also significantly contribute in plant tolerance to salt stress [104].

3.1.5 Microorganisms

Numerous plant growth-promoting microorganisms have been identified, which have become a useful tool for achieving sustainable agricultural production [105, 106]. Interestingly, rhizosphere is the home to numerous plant growth-promoting bacteria (PGPB) (or plant growth-promoting rhizobacteria; PGPR) (**Figure** 7). Most of these PGPB/PGPR are cheap and easily available resources; hence, they are generally used as an inoculant for bio-stimulation, biocontrol and biofertilization for the mitigation of salinity impacts in plants. Therefore, PGPB/PGPR has been argued as an alternative strategy for salt tolerance in plants [106–108]. The role of plant growth-promoting rhizobacteria (PGPR) in plant salinity tolerance and immunity is worth mentioning, where most PGPR protect plants by colonizing within the rhizosphere and producing antimicrobial metabolites (antagonistic) and producing regulatory hormones [109]. Plant inoculation with selected PGPRs (including *Bacillus pumilus* and *Pseudomonas pseudoalcaligenes*) under saline conditions resulted in increased uptake of N, P, K, decreased uptake of Na and Ca; and improved growth traits [110]. Thus, the use of selected PGPR makes the solubility and bioavailability of major mineral nutrients (such as soil-P) feasible under saline condition. PGPR strain-mediated induction of antioxidant enzymes can also be promising in improving plant salinity tolerance.

The role of arbuscular mycorrhizal fungi (AMF), a unique group of root obligate endophytic fungal symbionts (reported in about 90% of terrestrial plants), in plant salinity tolerance has also been found significant [17, 91]. AMF-mediated improvements in plant health involved enhanced ability of plants to replace K^+ with Na^+ using various transporters [111]; maintenance of high K^+ in roots and shoots [112, 113]; efficient regulation of K^+/Na^+ ratio required for cytoplasmic ion-balance maintenance [18]; strengthening of antioxidant defense mechanism [114]; improving water-use-efficiency, and compartmentalization of Na^+ within plant tissues

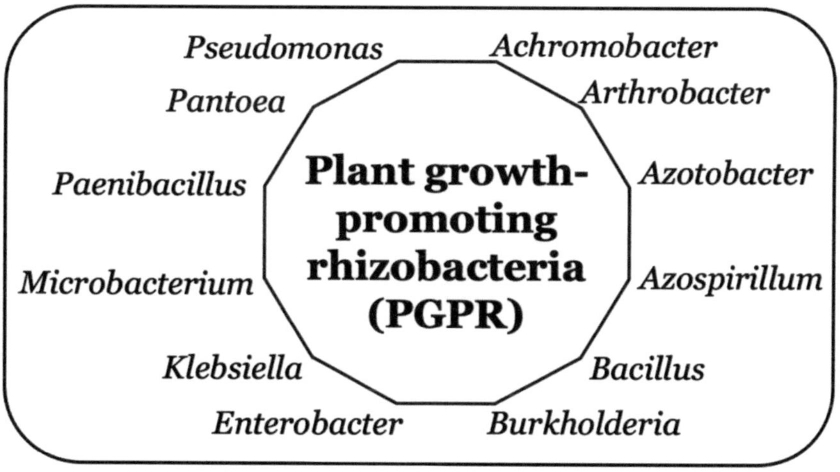

Figure 7.
Schematic representation of the major plant growth-promoting rhizobacteria inhabiting the plant rhizosphere [106–108].

[115, 116]; and inducing the osmolytic solutes (including proline, glycine betaine, or soluble sugars) [117].

3.1.6 Phenolic compounds

Elevated accumulation of phenolic compounds has been reported in salinity-impacted plants [118–120]. Considered significant among the most widely distributed secondary metabolites in the plant kingdom, phenolic compounds (including polyphenols and flavonoids) are the low-molecular-weight non-enzymatic antioxidants generated in plant cells. Most phenolic compounds act as signaling molecules and mediate auxin transport [121]. These phenolic compounds improve plant salinity tolerance via acting as reactive oxygen species (ROS)-scavengers, thereby maintaining a fine cellular redox homeostasis [118, 119].

3.1.7 Bio-stimulants

Bio-stimulants (also termed as bio-effectors) are viable microorganisms or active natural compounds, and can fall within four prime groups: acids (*humic acid*, *fulvic acid*, *humins*, *amino acids*, *fatty acids*, and *organic acids)*, microbes *(plant growth-promoting rhizobacteria*, *PGPR*; *arbuscular mycorrhizal fungi*, *AMF*; *Trichoderma* spp.), plant-derived bioactive substances *(polyphenols* and *allelochemicals, etc.*), and others (*beneficial elements: Al*, *Si*, *Na*, *Se*, *Co*, etc.). Most bio-stimulants are considered as the regulators of both ROS-metabolism and also stress metabolites involved in enhancing plant tolerance to major stresses (including salinity) [122, 123]. Bio-stimulants-mediated improved salinity tolerance in plants involved bio-stimulants-induced maintenance of reduced non-enzymatic antioxidants (such as GSH and AsA); improved tissue water status, ionic and nutrient homeostasis, and osmotic tolerance; stabilized membrane properties; maintenance of a fine-tuning among antioxidant enzymes; and decreased ROS generation electrolyte leakage and lipid peroxidation) [7, 124–126].

3.1.8 Nanomaterials

Nanomaterials (NMs) are materials with a basic structure of 1–100 nm in at least one dimension. Nanomaterials (in terms of nano-fertilizer and nano-pesticides, and other plant protection nano-agents) have shown great potential in agriculture [127]. The major mechanisms underlying NMs-mediated improved plant salinity tolerance included NMs-mediated alleviation of osmotic and ionic stress, enabling the better ability to maintain cytosolic K^+/Na^+ ratio; enhancing leaf mesophyll K^+ retention; efficient scavenging of ROS; maintaining a fine-tuning among the components of antioxidant defense system; and protection of photosynthesis [128–131]. The role of nanozymes in salinity-exposed plants has also been reported, where the supplied nanozymes with ROS-scavenging ability in plants helped to improve plant salt tolerance via maintaining ROS homeostasis and alleviating ROS-accumulation in plant organs [132, 133].

3.1.9 Molecular-genetic approach

Deep understanding of molecular insights into plant salinity stress tolerance has come to light mainly due to exhaustive studies on omics techniques. The list of the mentioned techniques comprises transcriptomics, genomics, proteomics, and metabolomics, where crucial cell signaling compounds crosstalk and integrative multi-omics techniques could be employed for improving salinity tolerance in plants [134, 135]. Notably, plants are of multi-genetic nature, which makes actual understanding of their responses to salinity very difficult. To this end, genomics (studies on a certain genome aimed at unveiling insights into organism's biology) has greatly helped identify and characterize the salinity stress response gene [135, 136]. Notably, understanding plants' salinity tolerance and also the development of salinity tolerant plants have been enabled to a great extent by employing high-throughput approaches including forward genetics, serial analysis of gene expression, expression sequence tag, next-generation sequencing), targeting-induced local lesion in genomes, RNA interference, and genome-wide association study [135, 137–139]. Additionally, plant salinity tolerance mechanisms are very complicated and involve polygenic traits [15]. To this end, dealing mainly with the RNA expression profile of organisms at temporal and spatial bases, transcriptomics has helped in the identification of transcripts/genes essential in controlling transcription and translation machinery in several studies on plants under salinity stress [20, 140–142].

Critical studies on the protein profiles (*which is actually the expression-reflection of the salinity-caused genes*) employing the proteomics approach have helped to identify proteins, and get their expression profile, post-translational modifications, and protein-protein interactions in both agricultural [143–145] and non-agricultural [146, 147] plants under salinity stress. Clear and reliable information about the major metabolites (such as most organic acids, hormones, amino acids, ketones, vitamins, and steroids) in salinity-exposed plants has been obtained employing metabolomics [47, 148, 149]. The collection of minerals and elements of an organism is considered as 'ionome', which has helped in understanding the role of adoption of controlled ion uptake, distribution system (homeostasis), and detoxification as plant's major strategy for adjusting high salinity [142, 143, 150]. The major impacts of salt stress on photosynthesis and related variables, ionic relationships, plant senescence, and yield can be assessed well by employing high-throughput phenotyping [151, 152].

The manipulation of N-dynamics through genetic engineering has great potential to improve plant life under salinity stress [153]. As also mentioned above that, ethylene is an important gaseous phytohormone involved in the regulation of plants growth, development, and senescence [154]. Among the small transcription factor gene families in higher plants, ethylene-insensitive 3 (EIN3)/ ethylene-insensitive 3-like (EIL) gene family is very important. All members of the EIN3/EIL gene family are key genes in the ethylene signaling pathway [155–157]. Ethylene-mediated downstream transcriptional cascade has been reported to involve EIN3/EILs as the major key elements and positive factors [158]. Ethylene-mediated enhanced salt tolerance in *Arabidopsis* involved EIN3/EIL1, which promoted EBF1/EBF2 proteasomal degradation and modulated several EIN3/ EIL1-regulated genes. Further, overexpression of EIN3 target genes (e.g., ERFs and SALT INDUCED EIN3/EIL1-DEPENDENT1 (SIED1)) can also improve plant salinity tolerance [159].

3.2 Saline soil health-management approaches

Appropriate, effective, cheap and environment-friendly management/amelioration approaches can be applied to improve the health (physicochemical and biological traits) of saline soils, and thereby providing the soil conditions favorable for establishment, and optimum growth and development of plants, which in turn can contribute in securing foods for future generations [160, 161]. Reclamation (salt removal from the plant-root zone) stands at the top of the saline soil health-management approaches, which is mainly comprised of reducing salinity by leaching; salt scraping; and phytoremediation of accumulated salt. Approaches such as the addition of organic amendments as ameliorant (such as crop residue, compost, farm yard manure, cattle manure, poultry manure, clover hay and wheat straw; aimed at improving soil physical conditions; stabilization of soil aggregates and improving water holding capacity); addition of chemical amendments (flushing out of the Na^+ from the root zone; can be done using gypsum, $CaSO_4$; lime, $CaCO_3$; sulfuric acid, H_2SO_4; hydrochloric acid, HCl; nitric acid, HNO_3); and soil ripping (reduce the compaction and assist with salt leaching; normally considered in sodic soils with compaction problem) may be adopted to reduce salt leaching from the upper layers of the soil. Salt scraping approach involves the physical removal of the salt crust, followed by leaching for salts (including Na^+) removal from the root zone [162, 163].

Phytoremediation is a plant-assisted approach alternative and efficient technique that has the potential to replace the abovementioned costly physical and chemical methods for reclaiming sodic and saline-sodic soils. Interestingly, the phytoremediation approach is based on the ability of plant roots to enhance the dissolution rate of native calcite and the removal of Na^+ [161, 164]. Plant species exhibiting hyperaccumulation of elevated soil salts and strong salinity tolerance can be considered in the phytoremediation-based management of saline soils. In particular, halophytes are the plants reported to exhibit salt resistance or salt tolerance to soils with >200 mM NaCl [165]. Notably, most halophytes (including grasses, shrubs, and trees) exhibit their extraordinary capacity of salt exclusion, excretion or salt accumulating at cellular, organelle and whole body levels. Given this, several halophytes have been considered a panacea in the remediation of salt-affected problematic soils (**Figure 8**). The use of phytoremediation approach

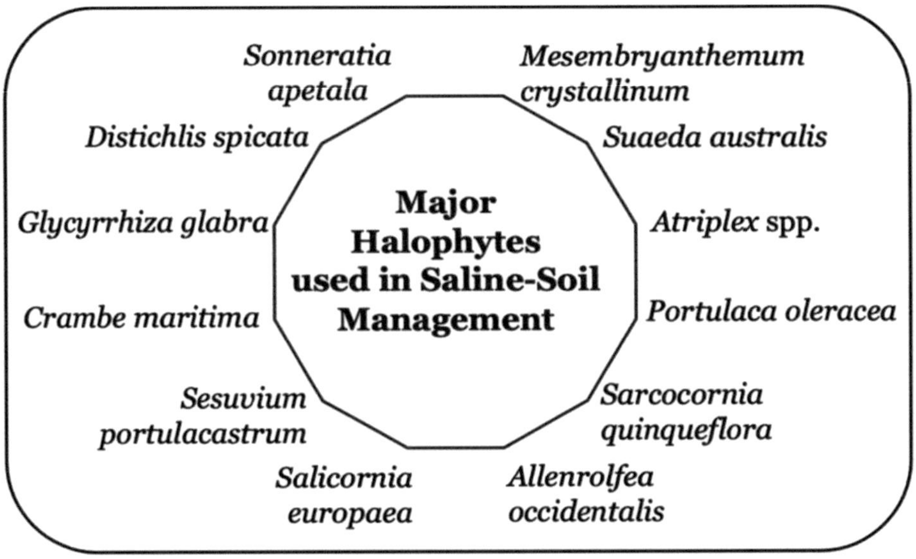

Figure 8.
Schematic representation of the major halophytes used in saline soil management [161, 164].

in salinity-loaded soils helped in improving physical, chemical, and microbiological properties; increasing soil fertility, organic matter accumulation, erodibility reduction, and increasing soil water retention [166–168].

4. Conclusions and prospects

This chapter overviewed the concept and the global status of soil salinization; highlighted the major mechanisms underlying salinity impacts in plants, and critically discussed in detail the potential approaches for making plant life easier under salinity. Notably, salinity-accrued decrease in plant-water status led to salinity-mediated impairments in growth and development; ionic imbalance, impaired nutrient uptake and assimilation were argued as a result of elevated accumulation of Na^+ and Cl^-; and impact on the tuning among the components of antioxidant defense system (causing oxidative stress) was cumulative resulted in severely hampered plant growth, metabolism, development and productivity. Notably, despite the fact of a very complex nature of plant salinity responses, the major approaches considered so far, have mainly focused on physiological/biochemical and agronomical (and molecular) aspects of salinity-exposed plants. Integrating these aspects with more molecular-genetic aspects, a critical crosstalk on gasotransmitters, phytohormones and mineral nutrients; intricacies therein of potential synergism and antagonism may help in understanding and getting insights into the complexity of signaling pathways. Evaluating how AMF influence the cell wall; and lipid metabolism under saline conditions will also be important.

Little success has been achieved in the field of breeding and genetic engineering of plants for their improved salinity tolerance. Exhaustive studies on the PGPR-colonization within the rhizosphere and the production of antimicrobial metabolites may also yield promising outcomes in plant salinity response and tolerance research. The use of varied plant bio-stimulants (bio-effectors) can also be employed to stimulate growth, nutrient uptake and crop plant salinity tolerance. Most mechanisms underlying the control of ion homeostasis, cell activity responses, and epigenetic regulation have been extensively unveiled in salinity-exposed. However, further insights into the identification of Na^+ sensor or receptor, ontogenic variation in the salt-induced signaling, and salt-tolerance markers for crop breeding can also be promising in future research on plant salinity stress tolerance. Additionally, further dissection and use of the outcomes of the crucial cell signaling compounds crosstalk and integrative multi-omics techniques will help further understand plants salinity responses and develop salinity tolerant plants.

Author details

Naser A. Anjum[1*], Asim Masood[1], Palaniswamy Thangavel[2] and Nafees A. Khan[1]

1 Department of Botany, Aligarh Muslim University, Aligarh, India

2 Department of Environmental Science, Periyar University, Salem, India

*Address all correspondence to: naanjum1@myamu.ac.in

References

[1] Singh RB. Climate change and food security. In: Tuteja N, Gill SS, Tuteja R, editors. Improving Crop Productivity in Sustainable Agriculture. Weinheim: Wiley-VCH Verlag; 2012. pp. 1-22

[2] Brevik EC. Soils and human health: An overview. In: Brevik EC, Burgess LC, editors. Soils and Human Health. Boca Raton, FL, USA: CRC Press; 2013. pp. 29-56

[3] Anjum NA, Gill SS, Gill R. Plant Adaptation to Environmental Change: Significance of Amino Acids and their Derivatives. 1st ed. Wallingford: CABI; 2014

[4] Anjum NA. Plant acclimation to environmental stress: A critical appraisal. Frontiers in Plant Science. 2015;**6**:445

[5] Lal R, Bouma J, Brevik E, Dawson L, Field DJ, Glaser B, et al. Soils and sustainable development goals of the United Nations: An International Union of Soil Sciences perspective. Geoderma Regional. 2021;**25**:e00398

[6] FAO-GloSIS. Global Soil Information System, FAO Global Soil Partnership. 2023. Available from: https://data.apps.fao.org/glosis/?lang=en [Accessed: 6 October 2023]

[7] Hassan FAS, Ali E, Gaber A, Fetouh MI, Mazrou R. Chitosan nanoparticles effectively combat salinity stress by enhancing antioxidant activity and alkaloid biosynthesis in *Catharanthus roseus* (L.) G. Don. Plant Physiology and Biochemistry. 2021;**162**:291-300

[8] Alexakis DD, Daliakopoulos IN, Panagea IS, Tsanis IK. Assessing soil salinity using WorldView-2 multispectral images in Timpaki, Crete, Greece. Geocarto International. 2018;**33**(4):321-338

[9] Osman KT. Saline and sodic soils. In: Management of Soil Problems. Cham: Springer; 2018

[10] Gorji T, Yildirim A, Hamzehpour N, Tanik A, Sertel E. Soil salinity analysis of Urmia Lake Basin using Landsat-8 OLI and sentinel-2A based spectral indices and electrical conductivity measurements. Ecological Indicators. 2020;**112**:106173

[11] Gharaibeh MA, Albalasmeh AA, El Hanandeh A. Estimation of saturated paste electrical conductivity using three modelling approaches: Traditional dilution extracts; saturation percentage and artificial neural networks. Catena. 2021;**200**:105141

[12] Omuto CT, Vargas RR, Elmobarak AA, Mapeshoane BE, Koetlisi KA, Ahmadzai H, et al. Digital soil assessment in support of a soil information system for monitoring salinization and sodification in agricultural areas. Land Degradation and Development. 2022;**33**(8):1204-1218

[13] Omuto CT, Scherstjanoi M, Kader MA, Musana B, Barman A, Fantappiè M, et al. Harmonization service and global library of models to support country-driven global information on salt-affected soils. Scientific Reports. 2023;**13**(1):13157

[14] Calone R, Bregaglio S, Sanoubar R, Noli E, Lambertini C, Barbanti L. Physiological adaptation to water salinity in six wild halophytes suitable for Mediterranean agriculture. Plants. 2021;**10**(2):309

[15] Munns R, Tester M. Mechanisms of salinity tolerance. Annual Review of Plant Biology. 2008;**59**:651-681

[16] Marschner H. Mineral Nutrition of Higher Plants. 3rd ed. London: Academic Press; 2012

[17] Thangavel P, Anjum NA, Muthukumar T, Sridevi G, Vasudhevan P, Maruthupandian A. Arbuscular mycorrhizae: Natural modulators of plant-nutrient relation and growth in stressful environments. Archives of Microbiology. 2022;**204**(5):264

[18] Borde M, Dudhane M, Kulkarni M. Role of arbuscular mycorrhizal fungi (AMF) in salinity tolerance and growth response in plants under salt stress conditions. In: Varma A, Prasad R, Tuteja N, editors. Mycorrhiza-Eco-Physiology, Secondary Metabolites, Nanomaterials. Cham: Springer; 2017. pp. 71-86

[19] Munns R. Genes and salt tolerance: Bringing them together. New Phytologist. 2005;**167**:645-663

[20] Jamil A, Riaz S, Ashraf M, Foolad MR. Gene expression profiling of plants under salt stress. Critical Reviews in Plant Sciences. 2011;**30**(5):435-458

[21] Rasheed F, Anjum NA, Masood A, Sofo A, Khan NA. The key roles of salicylic acid and sulfur in plant salinity stress tolerance. Journal of Plant Growth Regulation. 2022;**41**:1891-1904

[22] Jahan B, Iqbal N, Fatma M, Sehar Z, Masood A, Sofo A, et al. Ethylene supplementation combined with split application of nitrogen and sulfur protects salt-inhibited photosynthesis through optimization of proline metabolism and antioxidant system in mustard (*Brassica juncea* L.). Plants. 2021;**10**:1303

[23] Chowdhury FT, Halim MA, Hossain F, Akhtar N. Effects of sodium chloride on germination and seedling growth of sunflower (*Helianthus annuus* L.). Jahangirnagar University. Journal of Biological Sciences. 2018;7:35-44

[24] Sharma I, Ching E, Saini S, Bhardwaj R, Pati PK. Exogenous application of brassinosteroid offers tolerance to salinity by altering stress responses in rice variety Pusa Basmati-1. Plant Physiology and Biochemistry. 2013;**69**:17-26

[25] Rahneshan Z, Nasibi F, Moghadam AA. Effects of salinity stress on some growth, physiological, biochemical parameters and nutrients in two pistachio (*Pistacia vera* L.) rootstocks. Journal of Plant Interactions. 2018;**13**(1):73-82

[26] Giri B, Mukerji KG. Mycorrhizal inoculant alleviates salt stress in *Sesbania aegyptiaca* and Sesbania grandiflora under field conditions: Evidence for reduced sodium and improved magnesium uptake. Mycorrhiza. 2004;**14**:307-312

[27] Chaves MM, Flexas J, Pinheiro C. Photosynthesis under drought and salt stress: Regulation mechanisms from whole plant to cell. Annal of Botany. 2009;**103**:551-560

[28] Wani AS, Ahmad A, Hayat S, Tahir I. Epibrassinolide and proline alleviate the photosynthetic and yield inhibition under salt stress by acting on antioxidant system in mustard. Plant Physiology and Biochemistry. 2019;**135**:385-394

[29] Zörb C, Geilfus CM, Dietz KJ. Salinity and crop yield. Plant Biology. 2019;**21**:31-38

[30] Grattan SR, Grieve CM. Mineral element acquisition and growth response of plants grown in saline environments. Agriculture, Ecosystem and Environment. 1992;**38**:275-300

[31] Grattan SR, Grieve CM. Mineral nutrient acquisition and response by plants grown in saline environments. In: Pessarakli M, editor. Handbook of Plant and Crop Stress. 2nd ed. Basel, NY: CRC Press; 1999. pp. 203-229

[32] Azcon-G de Aguilar C, Azcón R, Barea JM. Endomycorrhizal fungi and rhizobium as biological fertilisers for *Medicago sativa* in normal cultivation. Nature. 1979;**279**:325-327

[33] Hodge A, Fitter AH. Substantial nitrogen acquisition by arbuscular mycorrhizal fungi from organic material has implications for N cycling. Proceedings of National Academy of Sciences USA. 2010;**107**:13754-13759

[34] Parihar P, Singh S, Singh R, Singh VP, Prasad SM. Effect of salinity stress on plants and its tolerance strategies: A review. Environmental Science and Pollution Research. 2015;**22**:4056-4075

[35] Wu H, Zhu M, Shabala L, Zhou M, Shabala S. K^+ retention in leaf mesophyll, an overlooked component of salinity tolerance mechanism: A case study for barley. Journal of Integrative Plant Biology. 2015;**57**(2):171-185

[36] Wu J, Shu S, Li C, Sun J, Guo S. Spermidine-mediated hydrogenperoxide signaling enhances the antioxidant capacity of salt-stressed cucumber roots. Plant Physiology and Biochemistry. 2018;**128**:152-162

[37] Fahad S, Hussain S, Saud S, Hassan S, Tanveer M, Ihsan MZ, et al. A combined application of biochar and phosphorus alleviates heat-induced adversities on physiological, agronomical and quality attributes of rice. Plant Physiology and Biochemistry. 2016;**103**:191-198

[38] Tabassum T, Farooq M, Ahmad R, Zohaib A, Wahid A. Seed priming and transgenerational drought memory improves tolerance against salt stress in bread wheat. Plant Physiology and Biochemistry. 2017;**118**:362-369

[39] Astolfi S, Zuchi S. Adequate S supply protects barley plants from adverse effects of salinity stress by increasing thiol contents. Acta Physiologia Plantarum. 2013;**35**:175-181

[40] Shao QS, Shu S, Du J, Xing WW, Guo SR, Sun J. Effects of NaCl stress on nitrogen metabolism of cucumber seedlings. Russian Journal of Plant Physiology. 2015;**62**:595-603

[41] Fuertes-Mendizábal T, Bastías EI, González-Murua C, González-Moro M. Nitrogen assimilation in the highly salt and boron-tolerant ecotype *Zea mays* L. Amylacea. Plants. 2020;**9**:322

[42] Goel P, Singh AK. Abiotic stresses downregulate key genes involved in nitrogen uptake and assimilation in *Brassica juncea* L. PLoS One. 2015;**10**:e0143645

[43] Tzortzakis N, Pitsikoulaki G, Stamatakis A, Chrysargyris A. Ammonium to total nitrogen ratio interactive effects with salinity application on *Solanum lycopersicum* growth, physiology, and fruit storage in a closed hydroponic system. Agronomy. 2022;**12**(2):386

[44] Dalio RJD, Pinheiro HP, Sodek L, Haddad CRB. 24-epibrassinolide restores nitrogen metabolism of pigeon pea under saline stress. Botanical Studies. 2013;**54**(1):1-7

[45] Shahzad R, Khan AL, Bilal S, Waqas M, Kang SM, Lee IJ. Inoculation of abscisic acid-producing endophytic bacteria enhances salinity stress tolerance in Oryza sativa. Environmental and Experimental Botany. 2017;**136**:68-77

[46] Füzy A, Biró B, Tóth T, Hildebrandt U, Bothe H. Drought, but not salinity, determines the apparent effectiveness of halophytes colonized by arbuscular mycorrhizal fungi. Journal of Plant Physiology. 2008;**165**:1181-1192

[47] AbdElgawad H, Zinta G, Hegab MM, Pandey R, Asard H, Abuelsoud W. High salinity induces different oxidative stress and antioxidant responses in maize seedlings organs. Frontiers in Plant Science. 2016;7:276

[48] Sehar Z, Masood A, Khan NA. Nitric oxide reverses glucose-mediated photosynthetic repression in wheat (*Triticum aestivum* L.) under salt stress. Environmental and Experimental Botany. 2019;**161**:277-289

[49] Anjum NA, Umar S, Chan MT. Ascorbate-Glutathione Pathway and Stress Tolerance in Plants. Dordrecht: Springer; 2010

[50] Anjum NA, Ahmad I, Mohmood I, Pacheco M, Duarte AC, Pereira E, et al. Modulation of glutathione and its related enzymes in plants' responses to toxic metals and metalloids—A review. Environmental and Experimental Botany. 2012;**75**:307-324

[51] Li Z, Zhu L, Zhao F, Li J, Zhang X, Kong X, et al. Plant salinity stress response and nano-enabled plant salt tolerance. Frontiers in Plant Science. 2022;**13**:843994

[52] Yang Y, Guo Y. Elucidating the molecular mechanisms mediating plant salt-stress responses. New Phytologist. 2018;**217**(2):523-539

[53] Ahanger MA, Qin C, Begum N, Maodong Q, Dong XX, El-Esawi M, et al. Nitrogen availability prevents oxidative effects of salinity on wheat growth and photosynthesis by up-regulating the antioxidants and osmolytes metabolism; and secondary metabolite accumulation. BMC Plant Biolology. 2019;**19**:479

[54] Singh A, Hussain I, Singh NB, Singh H. Uptake, translocation and impact of green synthesized nanoceria on growth and antioxidant enzymes activity of *Solanum lycopersicum* L. Ecotoxicology and Environmental Safety. 2019;**182**:109410

[55] Soliman M, Elkelish A, Souad T, Alhaithloul H, Farooq M. Brassinosteroid seed priming with nitrogen supplementation improves salt tolerance in soybean. Physiology and Molecular Biology of Plants. 2020;**26**:501-511

[56] Per TS, Khan NA, Masood A, Fatma M. Methyl jasmonate alleviates cadmium-induced photosynthetic damages through increased S-assimilation and glutathione production in mustard. Frontiers in Plant Science. 2016;**7**:1933

[57] Fatma M, Masood A, Per TS, Rasheed F, Khan NA. Interplay between nitric oxide and sulfur assimilation in salt tolerance in plants. Crop Journal. 2016;**4**:153-161

[58] Sripinyowanich S, Klomsakul P, Boonburapong B, Bangyeekhun T, Asami T, et al. Exogenous ABA induces salt tolerance in Indica rice (*Oryza sativa* L.): The role of OsP5CS1 and OsP5CR gene expression during salt stress. Environmental and Experimental Botany. 2013;**86**:94-105

[59] Ruggiero A, Landi S, Punzo P, Possenti M, Van Oosten MJ, Costa A, et al. Salinity and ABA seed responses in pepper: Expression and interaction of ABA core signaling components. Frontiers in Plant Science. 2019;**10**:304

[60] Majid A, Rather BA, Masood A, Sehar Z, Anjum NA, Khan NA. Abscisic

acid in coordination with nitrogen alleviates salinity-inhibited photosynthetic potential in mustard by improving proline accumulation and antioxidant activity. Stress. 2021;**1**(3):162-180

[61] Iglesias MJ, Terrile MC, Windels D, Lombardo MC, Bartoli CG, Vazquez F, et al. *MiR393* regulation of auxin signaling and redox-related components during acclimation to salinity in *Arabidopsis*. PLoS One. 2014;**9**:e107678

[62] Liu W, Li RJ, Han TT, Cai W, Fu ZW, Lu YT. Salt stress reduces root meristem size by nitric oxide-mediated modulation of auxin accumulation and signaling in *Arabidopsis*. Plant Physiology. 2015;**168**:343-356

[63] Ryu H, Cho YG. Plant hormones in salt stress tolerance. Journal of Plant Biology. 2015;**58**:147-155

[64] Ding HD, Zhu XH, Zhu ZW, Yang SJ, Zha DS, et al. Amelioration of salt-induced oxidative stress in eggplant by application of 24-epibrassinolide. Biologia Plantarum. 2012;**56**:767-770

[65] Fariduddin Q, Mir BA, Yusuf M, Ahmad A. Comparative roles of brassinosteroids and polyamines in salt stress tolerance. Acta Physiologia Plantarum. 2013;**35**:2037-2053

[66] Azzam CR, Zaki SN, Bamagoos AA, Rady MM, Alharby HF. Soaking maize seeds in zeatin-type cytokinin biostimulators improves salt tolerance by enhancing the antioxidant system and photosynthetic efficiency. Plants. 2022;**11**(8):1004

[67] Yin Y, Yang T, Li S, Li X, Wang W, Fan S. Transcriptomic analysis reveals that methyl jasmonate confers salt tolerance in alfalfa by regulating antioxidant activity and ion homeostasis. Frontiers in Plant Science. 2023;**14**:1258498

[68] Iqbal N, Khan NA, Ferrante A, Trivellini A, Francini A, Khan MI. Ethylene role in plant growth, development and senescence: Interaction with other phytohormones. Frontiers in Plant Science. 2017;**8**:475

[69] Al Murad M, Razi K, Benjamin LK, Lee JH, Kim TH, Muneer S. Ethylene regulates sulfur acquisition by regulating the expression of sulfate transporter genes in oilseed rape. Physiologia Plantarum. 2020;**171**:533-545

[70] Borbély P, Poór P, Tari I. Changes in physiological and photosynthetic parameters in tomato of different ethylene status under salt stress: Effects of exogenous 1-aminocyclopropane-1-carboxylic acid treatment and the inhibition of ethylene signalling. Plant Physiology and Biochemistry. 2020;**156**:345-356

[71] Iqbal H, Yaning C, Waqas M, Rehman HM, Sharee FM, Iqbal S. Hydrogen peroxide application improves quinoa performance by affecting physiological and biochemical mechanisms under water-deficit conditions. Journal of Agronomy and Crop Sciences. 2018;**204**:541-553

[72] Rady MM, Talaat NB, Abdelhamid MT, Shawky BT, Desoky EM. Maize (*Zea mays* L.) grains extract mitigates the deleterious effects of salt stress on common bean (*Phaseolus vulgaris* L.) growth and physiology. Journal of Horticultural Science and Biotechnology. 2021;**94**:777-789

[73] Iftikhar A, Ali S, Yasmeen T, Arif MS, Zubair M, Rizwan M, et al. Effect of gibberellic acid on growth, photosynthesis and antioxidant defense

system of wheat under zinc oxide nanoparticle stress. Environmental Pollution. 2019;**254**:113109

[74] Taheri Z, Vatankhah E, Jafarian V. Methyl jasmonate improves physiological and biochemical responses of *Anchusa italica* under salinity stress. South African Journal of Botany. 2020;**130**:375-382

[75] Sheteiwy MS, Ulhassan Z, Qi W, Lu H, AbdElgawad H, Minkina T, et al. Association of jasmonic acid priming with multiple defense mechanisms in wheat plants under high salt stress. Frontiers in Plant Science. 2022;**13**:886862

[76] Fancy NN, Bahlmann AK, Loake GJ. Nitric oxide functions in plant abiotic stress. Plant, Cell and Environment. 2017;**40**:462-472

[77] Corpas FJ, González-Gordo S, Cañas A, Palma JM. Nitric oxide and hydrogen sulfide in plants: Which comes first? Journal of Experimental Botany. 2019;**70**(17):4391-4404

[78] Arora D, Bhatla SC. Melatonin and nitric oxide regulate sunflower seedling growth under salt stress accompanying differential expression of Cu/Zn SOD and Mn SOD. Free Radical Biology and Medicine. 2017;**106**:315-328

[79] Gadelha CG, de Souza MR, Alencar NLM, Costa JH, Prisco JT, Gomes-Filho E. Exogenous nitric oxide improves salt tolerance during establishment of *Jatropha curcas* seedlings by ameliorating oxidative damage and toxic ion accumulation. Journal of Plant Physiology. 2017;**212**:69-79

[80] Poór P, Tari I. Ethylene-regulated reactive oxygen species and nitric oxide under salt stress in tomato cell suspension culture. Acta Biologica Szegediensis. 2011;**55**(1):143-146

[81] Zhao G, Zhao Y, Yu X, Kiprotich F, Han H, Guan R, et al. Nitric oxide is required for melatonin-enhanced tolerance against salinity stress in rapeseed (*Brassica napus* L.) seedlings. International Journal of Molecular Sciences. 2018;**19**(7):1912

[82] Csiszár J, Horváth E, Váry Z, Gallé Á, Bela K, Brunner S, et al. Glutathione transferase supergene family in tomato: Salt stress-regulated expression of representative genes from distinct GST classes in plants primed with salicylic acid. Plant Physiology and Biochemistry. 2014;**78**:15-26

[83] Asensi-Fabado MA, Munné-Bosch S. The aba3-1 mutant of *Arabidopsis thaliana* withstands moderate doses of salt stress by modulating leaf growth and salicylic acid levels. Journal of Plant Growth Regulation. 2011;**30**:456-466

[84] Sousa VF, Santos AS, Sales WS, Silva AJ, Gomes FA, Dias TJ, et al. Exogenous application of salicylic acid induces salinity tolerance in eggplant seedlings. Brazilian Journal of Biology. 2022;**84**:e257739

[85] Aroca R, Ruiz-Lozano JM, Zamarreño ÁM, Paz JA, García-Mina JM, Pozo MJ, et al. Arbuscular mycorrhizal symbiosis influences strigolactone production under salinity and alleviates salt stress in lettuce plants. Journal of Plant Physiology. 2013;**170**:47-55

[86] Ma N, Hu C, Wan L, Hu Q, Xiong J, Zhang C. Strigolactones improve plant growth, photosynthesis, and alleviate oxidative stress under salinity in rapeseed (*Brassica napus* L.) by regulating gene expression. Frontiers in Plant Science. 2017;**8**:1671

[87] Ren CG, Kong CC, Xie ZH. Role of abscisic acid in strigolactone-induced salt stress tolerance in arbuscular

mycorrhizal *Sesbania cannabina* seedlings. BMC Plant Biology. 2018;**18**:74

[88] Yu Z, Duan X, Luo L, Dai S, Ding Z, Xia G. How plant hormones mediate salt stress responses. Trends in Plant Science. 2020;**25**(11):1117-1130

[89] Chen TH, Murata N. Glycinebetaine protects plants against abiotic stress: Mechanisms and biotechnological applications. Plant, Cell & Environment. 2011;**34**(1):1-20

[90] Singh M, Kumar J, Singh S, Singh VP, Prasad SM. Roles of osmoprotectants in improving salinity and drought tolerance in plants: A review. Reviews in Environmental Science and Bio/Technology. 2015;**14**:407-426

[91] Evelin H, Devi TS, Gupta S, Kapoor R. Mitigation of salinity stress in plants by arbuscular mycorrhizal symbiosis: Current understanding and new challenges. Frontiers in Plant Science. 2019;**10**:470

[92] Anjum NA, Thangavel P, Rasheed F, Masood A, Pirasteh-Anosheh H, Khan NA. Osmolytes: Efficient oxidative stress-busters in plants. In: Global Climate Change and Plant Stress Management. New Jersey, USA: John Wiley & Sons Ltd.; 2023. pp. 399-409

[93] Minocha R, Majumdar R, Minocha SC. Polyamines and biotic stress in plants: A complex relationship. Frontiers in Plant Science. 2014;**5**:175

[94] Zapata PJ, Serrano M, García-Legaz MF, Pretel MT, Botella MA. Short term effect of salt shock on ethylene and polyamines depends on plant salt sensitivity. Frontiers in Plant Science. 2017;**8**:855

[95] Tiburcio AF, Altabella T, Bitrián M, Alcázar R. The roles of polyamines during the lifespan of plants: From development to stress. Planta. 2014;**240**:1-18

[96] Parvin S, Lee OR, Sathiyaraj G, Khorolragchaa A, Kim YJ, Yang DC. Spermidine alleviates the growth of saline-stressed ginseng seedlings through antioxidative defense system. Gene. 2014;**537**(1):70-78

[97] Rathinapriya P, Pandian S, Rakkammal K, Balasangeetha M, Alexpandi R, Satish L, et al. The protective effects of polyamines on salinity stress tolerance in foxtail millet (*Setaria italica* L.), an important C4 model crop. Physiology and Molecular Biology of Plants. 2020;**26**:1815-1829

[98] Kaur G, Asthir B. Proline: A key player in plant abiotic stress tolerance. Biologia Plantarum. 2015;**59**:609-619

[99] Parvaiz A, Satyawati S. Salt stress and phyto-biochemical responses of plants-a review. Plant, Soil and Environment. 2008;**54**:89-99

[100] Lunn JE, Delorge I, Figueroa CM, Van Dijck P, Stitt M. Trehalose metabolism in plants. The Plant Journal. 2014;**79**(4):544-677

[101] Garg AK, Kim JK, Owens TG, Ranwala AP, Do Choi Y, Kochian LV, et al. Trehalose accumulation in rice plants confers high tolerance levels to different abiotic stresses. Proceedings of National Academy of Science USA. 2002;**99**:15898-15903

[102] Redillas MC, Park SH, Lee JW, Kim YS, Jeong JS, Jung H, et al. Accumulation of trehalose increases soluble sugar contents in rice plants conferring tolerance to drought and salt stress. Plant Biotechnology Reports. 2012;**6**:89-96

[103] Chang B, Yang L, Cong W, Zu Y, Tang Z. The improved resistance to high salinity induced by trehalose is associated with ionic regulation and osmotic adjustment in *Catharanthus roseus*. Plant Physiology and Biochemistry. 2014;**77**:140-148

[104] Guo LQ, Shi DC, Wang DL. The key physiological response to alkali stress by the alkali-resistant halophyte *Puccinellia tenuiflora* is the accumulation of large quantities of organic acids and into the rhyzosphere. Journal of Agronomy and Crop Science. 2010;**196**:123-135

[105] Rigobelo EC, Kandasamy S, Saravanakumar D. Editorial: Plant growth-promoting microorganisms for sustainable agricultural production. Frontiers in Sustainable Food System. 2022;**6**:842533

[106] Hemati A, Shafea L, Lajayer BA, Ghorbanpour M, Astatkie T. An overview of bacterial bio-fertilizers function on soil fertility under abiotic stresses. In: Plant Stress Mitigators: Types, Techniques and Functions. 2023. pp. 505-512

[107] Numan M, Bashir S, Khan Y, Mumtaz R, Shinwari ZK, Khan AL, et al. Plant growth promoting bacteria as an alternative strategy for salt tolerance in plants: A review. Microbiological Research. 2018;**209**:21-32

[108] Goswami M, Suresh DE. Plant growth-promoting rhizobacteria - alleviators of abiotic stresses in soil: A review. Pedosphere. 2020;**30**(1):40-61

[109] Maheshwari DK, Dheeman S, Agarwal M. Phytohormone-producing PGPR for sustainable agriculture. In: Maheshwari DK, editor. Bacterial Metabolites in Sustainable Agroecosystem. Cham: Springer; 2015. pp. 159-182

[110] Jha Y, Subramanian RB. Paddy plants inoculated with PGPR show better growth physiology and nutrient content under saline conditions. Chilean Journal Agricultural Research. 2013;**73**:2

[111] Saxena B, Shukla K, Giri B. Arbuscular mycorrhizal fungi and tolerance of salt stress in plants. In: Wu QS, editor. Arbuscular Mycorrhizas and Stress Tolerance of Plants. Singapore: Springer; 2017. pp. 67-97

[112] Giri B, Kapoor R, Mukerji KG. Improved tolerance of *Acacia nilotica* to salt stress by arbuscular mycorrhiza, *Glomus fasciculatum* may be partly related to elevated K/Na ratios in root and shoot tissues. Microbial Ecology. 2007;**54**:753-760

[113] Wu QS, Zou YN, He XH. Contributions of arbuscular mycorrhizal fungi to growth, photosynthesis, root morphology and ionic balance of citrus seedlings under salt stress. Acta Physiologia Plantarum. 2010;**32**:297-304

[114] Abdel Latef AAH, Hashem A, Rasool S, Abd-Allah EF, Alqarawi AA, Egamberdieva D, et al. Arbuscular mycorrhizal symbiosis and abiotic stress in plants: A review. Journal of Plant Biology. 2016;**59**:407-426

[115] Augé RM, Toler HD, Saxton AM. Arbuscular mycorrhizal symbiosis and osmotic adjustment in response to NaCl stress: A meta-analysis. Frontiers in Plant Science. 2014;**5**:562

[116] Wang H, An T, Huang D, Liu R, Xu B, Zhang S, et al. Arbuscular mycorrhizal symbioses alleviating salt stress in maize is associated with a decline in root-to-leaf gradient of Na^+/K^+ ratio. BMC Plant Biology. 2021;**21**:457

[117] Porcel R, Aroca R, Ruiz-Lozano JM. Salinity stress

alleviation using arbuscular mycorrhizal fungi. A review. Agronomy and Sustainable Development. 2012;**32**:181-200

[118] Trchounian A, Petrosyan M, Sahakyan N. Plant cell redox homeostasis and reactive oxygen species. In: Gupta D, Palma J, Corpas F, editors. Redox State as a Central Regulator of Plant-Cell Stress Responses. Cham: Springer; 2016

[119] Sirin S, Aslım B. Determination of antioxidant capacity, phenolic acid composition and antiproliferative effect associated with phenylalanine ammonia lyase (PAL) activity in some plants naturally growing under salt stress. Medicinal Chemistry Research. 2019;**28**:229-238

[120] Kiani R, Arzani A, Mirmohammady Maibody SA. Polyphenols, flavonoids, and antioxidant activity involved in salt tolerance in wheat, *Aegilops cylindrica* and their amphidiploids. Frontiers in Plant Science. 2021;**12**:646221

[121] Tohidi B, Rahimmalek M, Arzani A. Essential oil composition, total phenolic, flavonoid contents, and antioxidant activity of *Thymus* species collected from different regions of Iran. Food Chemistry. 2017;**220**:153-161

[122] Hasanuzzaman M, Parvin K, Bardhan K, Nahar K, Anee TI, Masud AA, et al. Biostimulants for the regulation of reactive oxygen species metabolism in plants under abiotic stress. Cell. 2021;**10**(10):2537

[123] Nawaz F, Majeed S, Farman M, Sheteiwy MS, Al-Mamun A, Nawaz M, et al. Biostimulants as regulators of stress metabolites to enhance drought and salinity stress tolerance in plants. In: Biostimulants for Crop Production and Sustainable Agriculture. GB: CABI; 2022. pp. 265-294

[124] Ait-El-Mokhtar M, Baslam M, Ben-Laouane R, Anli M, Boutasknit A, Mitsui T, et al. Alleviation of detrimental effects of salt stress on date palm (*Phoenix dactylifera* L.) by the application of arbuscular mycorrhizal fungi and/or compost. Frontiers in Sustainable Food Systetm. 2020;**4**:131

[125] Parvin K, Nahar K, Hasanuzzaman M, Bhuyan MB, Mohsin SM, Fujita M. Exogenous vanillic acid enhances salt tolerance of tomato: Insight into plant antioxidant defense and glyoxalase systems. Plant Physiology and Biochemistry. 2020;**150**:109-120

[126] Rehman H, Alharby HF, Bamagoos AA, Abdelhamid MT, Rady MM. Sequenced application of glutathione as an antioxidant with an organic biostimulant improves physiological and metabolic adaptation to salinity in wheat. Plant Physiology and Biochemistry. 2021;**158**:43-52

[127] Gade A, Ingle P, Nimbalkar U, Rai M, Raut R, Vedpathak M, et al. Nanofertilizers: The next generation of agrochemicals for Long-term impact on sustainability in farming systems. Agrochemicals. 2023;**2**(2):257-278

[128] Gao L, Zhuang J, Nie L, Zhang J, Zhang Y, Gu N, et al. Intrinsic peroxidase-like activity of ferromagnetic nanoparticles. Nature Nanotechnology. 2007;**2**:577-583

[129] Rico CM, Peralta-Videa JR, Gardea-Torresdey JL. Chemistry, biochememistry of nanoparticles, and their role in antioxidant defense system in plants. In: Siddiqui M, Al-Whaibi M, Mohammad F, editors. Nanotechnology and Plant Sciences. Cham: Springer; 2015. pp. 1-17

[130] Khan MN, Mobin M, Abbas ZK, AlMutairi KA, Siddiqui ZH. Role

of nanomaterials in plants under challenging environments. Plant Physiology and Biochemistry. 2017;**110**:194-209

[131] Pinedo-GuerreroZH,Cadenas-PliegoG, Ortega-Ortiz H, Gonzalez-Morales S, Benavideds-Mendoza A, et al. Form of silica improves yield, fruit quality and antioxidant defense system of tomato plants under salt stress. Agriculture. 2020;**10**:1-21

[132] An J, Hu P, Li F, Wu H, Shen Y, White JC, et al. Emerging investigator series: Molecular mechanisms of plant salinity stress tolerance improvement by seed priming with cerium oxide nanoparticles. Environmental Science - Nano. 2020;7:2214-2228

[133] Zhao L, Lu L, Wang A, Zhang H, Huang M, Wu H, et al. Nano-biotechnology in agriculture: Use of nanomaterials to promote plant growth and stress tolerance. Journal of Agricultural and Food Chemistry. 2020;**68**(7):1935-1947

[134] Kumari A, Das P, Parida AK, Agarwal PK. Proteomics, metabolomics, and ionomics perspectives of salinity tolerance in halophytes. Frontiers in Plant Science. 2015;**6**:537

[135] Singhal RK, Saha D, Skalicky M, Mishra UN, Chauhan J, Behera LP, et al. Crucial cell signaling compounds crosstalk and integrative multi-omics techniques for salinity stress tolerance in plants. Frontiers in Plant Science. 2021;**12**:670369

[136] Gilliham M, Able JA, Roy SJ. Translating knowledge about abiotic stress tolerance to breeding programmes. Plant Journal. 2017;**90**:898-917

[137] Tang RJ, Liu H, Bao Y, Lv QD, Yang L, Zhang HX. The woody plant poplar has a functionally conserved salt overly sensitive pathway in response to salinity stress. Plant Molecular Biology. 2010;**74**:367-380

[138] Kumar V, Singh A, Mithra SA, Krishnamurthy SL, Parida SK, Jain S, et al. Genome-wide association mapping of salinity tolerance in rice (*Oryza sativa*). DNA Research. 2015;**22**(2):133-145

[139] Ravelombola W, Shi A, Weng Y, Mou B, Motes D, Clark J, et al. Association analysis of salt tolerance in cowpea (*Vigna unguiculata* (L.) Walp) at germination and seedling stages. Theoretical and Applied Genetics. 2018;**131**:79-91

[140] Sahi C, Singh A, Kumar K, Blumwald E, Grover A. Salt stress response in rice: Genetics, molecular biology, and comparative genomics. Functional and Integrative Genomics. 2006;**6**:263-284

[141] Garg N, Singla P. Stimulation of nitrogen fixation and trehalose biosynthesis by naringenin (Nar) and arbuscular mycorrhiza (AM) in chickpea under salinity stress. Plant Growth Regulation. 2016;**80**:5-22

[142] Arshad M, Gruber MY, Wall K, Hannoufa A. An insight into *microRNA156* role in salinity stress responses of alfalfa. Frontiers in Plant Science. 2017;**8**:356

[143] Du CX, Fan HF, Guo SR, Tezuka T, Li J. Proteomic analysis of cucumber seedling roots subjected to salt stress. Phytochemistry. 2010;**71**:1450-1459

[144] Zörb C, Schmitt S, Mühling KH. Proteomic changes in maize roots after short–term adjustment to saline growth conditions. Proteomics. 2010;**10**:4441-4449

[145] Ngara R, Ndimba R, Borch-Jensen J, Jensen ON, Ndimba B. Identification and profiling of salinity stress-responsive proteins in Sorghum bicolor seedlings. Journal of Proteomics. 2012;**75**(13):4139-4150

[146] Razavizadeh R, Ehsanpour AA, Ahsan N, Komatsu S. Proteome analysis of tobacco leaves under salt stress. Peptides. 2009;**30**(9):1651-1659

[147] Chen S, Jiang J, Li H, Liu G. The salt-responsive transcriptome of *Populus simonii*× *Populus nigra* via DGE. Gene. 2012;**504**:203-212

[148] Lu Y, Lam H, Pi E, Zhan Q, Tsai S, Wang C, et al. Comparative metabolomics in *Glycine max* and *Glycine soja* under salt stress to reveal the phenotypes of their offspring. Journal of Agricultural and Food Chemistry. 2013;**61**(36):8711-8721

[149] Pan J, Li Z, Dai S, Ding H, Wang Q, Li X, et al. Integrative analyses of transcriptomics and metabolomics upon seed germination of foxtail millet in response to salinity. Scientific Reports. 2020;**10**(1):13660

[150] Sanadhya P, Agarwal P, Agarwal PK. Ion homeostasis in a salt-secreting halophytic grass. AoB Plants. 2015;7:1

[151] Humplík JF, Lazár D, Husičková A, Spíchal L. Automated phenotyping of plant shoots using imaging methods for analysis of plant stress responses—A review. Plant Methods. 2015;**11**:1-10

[152] Atieno J, Li Y, Langridge P, Dowling K, Brien C, Berger B, et al. Exploring genetic variation for salinity tolerance in chickpea using image-based phenotyping. Scientific Reports. 2017;7:1-11

[153] Nazir F, Mahajan M, Khatoon S, Albaqami M, Ashfaque F, Chhillar H, et al. Sustaining nitrogen dynamics: A critical aspect for improving salt tolerance in plants. Frontiers in Plant Science. 2023;**14**:1087946

[154] Abeles FB, Morgan PW, Saltveit ME Jr. Ethylene in Plant Biology. 2nd ed. San Diego: Academic Press; 1992

[155] Chen YF, Etheridge N, Schaller GE. Ethylene signal transduction. Annals of Botany. 2005;**95**:901-915

[156] Lin Z, Zhong S, Grierson D. Recent advances in ethylene research. Journal of Experimental Botany. 2009;**60**:3311-3336

[157] Li M, Wang R, Liang Z, et al. Genome-wide identification and analysis of the EIN3/EIL gene family in allotetraploid *Brassica napus* reveal its potential advantages during polyploidization. BMC Plant Biology. 2019;**19**:110

[158] Liu C, Li J, Zhu P, Yu J, Hou J, Wang C, et al. Mulberry EIL3 confers salt and drought tolerances and modulates ethylene biosynthetic gene expression. Peer Journal. 2019;7:e6391

[159] Zhang X, Shi Z, Tian Y, Zhou Q, Cai J, Dai T, et al. Salt stress increases content and size of glutenin macropolymers in wheat grain. Food Chemistry. 2016;**197**:516-521

[160] dos Santos MA, Freire MBGS, Freire FJ, da Rocha AT, de Lucena PG, et al. Reclamation of saline soil under association between *Atriplex nummularia* L. and *Glycophytes* plants. Agriculture. 2022;**12**:1124

[161] Srivastava N. Reclamation of saline and sodic soil through phytoremediation. In: Environmental Concerns and Sustainable Development, Biodiversity, Soil and Waste Management. Vol. 2. Singapore: Springer; 2020. pp. 279-306

[162] Yuvaraj M, Bose KS, Elavarasi P, Tawfik E. Soil Salinity and its Management. London, UK: IntechOpen; 2021

[163] Shaygan M, Baumgartl T. Reclamation of salt-affected land: A review. Soil Systems. 2022;**6**(3):61

[164] Hasanuzzaman M, Nahar K, Alam MM, Bhowmik PC, Hossain MA, Rahman MM, et al. Potential use of halophytes to remediate saline soils. Biomed Research International. 2014;**2014**:589341

[165] Meng X, Zhou J, Sui N. Mechanisms of salt tolerance in halophytes: Current understanding and recent advances. Open Life Science. 2018;**13**:149-154

[166] Zhang JB, Yang JS, Yao RJ, Yu SP, Li FR, Hou XJ. The effects of farmyard manure and mulch on soil physical properties in a reclaimed coastal tidal flat salt-affected soil. Journal of Integrative Agriculture. 2014;**13**:1782-1790

[167] Nouri H, Borujeni SC, Nirola R, Hassanli A, Beecham S, Alaghmand S, et al. Application of green remediation on soil salinity treatment: A review on halophytoremediation. Process Safety and Environmental Protection. 2017;**107**:94-107

[168] Leite MC, dos Santos Freire MB, de Queiroz JV, Maia LC, Duda GP, de Medeiros EV. Mycorrhizal *Atriplex nummularia* promote revegetation and shifts in microbial properties in saline Brazilian soil. Applied Soil Ecology. 2020;**153**:103574

Chapter 2

The Scale and Complexity of Salinity Impacts on Sri Lankan Rice Farming Systems: Actionable Insights

Bhagya Indeevari Dasanayaka,
Demuni Sumith de Zoysa Abeysiriwardena and
Chandima Kumudini Ariyarathna Hanchaplola Appuhamilage

Abstract

Saline-affected rice (*Oryza sativa* L.) production environments in Sri Lanka can be divided into three categories including: late-season salinity in irrigated mega-cultivation environments during the minor cultivation season where soil EC $\geq$7 dSm^{-1}, late season salinity in rain-fed farming systems in the west, southwest, and eastern coastal line during the minor cultivation season where soil EC $\geq$20 dSm^{-1}, and early season salinity in selected irrigated and rain fed sites during major and minor cultivation seasons as a result of residual overload of salts that was not washed off due to inadequate rain. In the west and southern coast early season salinity salinity can exceed EC $\geq$12 dSm^{-1}. The proposed zones of saline-afflicted production environments permit designing of target ideotypes and locally adapted rice varieties. Accordingly, high yielding, 3 to 3.5 months duration varieties that are tolerant at >7 dSm^{-1} are recommended for intensive irrigated farming systems affected due to late season salinity (panicle initiation stage of the crop, PI); high yielding, 2.5 to 3 months duration varieties can avoid late season salinity in intensive irrigated farming, and varieties tolerant up to EC = 12 to 20 dSm^{-1} throughout the crop life including seedling and PI stages can target saline affected, semi-subsistence rice cultivation in rain-fed systems. In fact, secondary salinization in local rice farming environments is resulting from interaction among multiple factors; therefore, system-level interventions are necessary to manage the impacts.

Keywords: secondary salinization, rice (*Oryza sativa* L.), rice cultivation seasons, saline tolerant germplasm, system interventions

1. Introduction

1.1 Background

Rice (*Oryza sativa*) is the staple food of an estimated 3.5 billion people from East and South Asia, the Middle East, the West Indies, and Latin America, accounting for 50–80% of their daily calorie intake [1]. It is also the primary source of income and employment for more than 200 million households in developing countries [2]. Rice is the main staple providing over 45% of the total caloric and 40% of the protein requirements for the Sri Lankan population [3, 4]. In the country, rice cultivation occupies 16% of the total land area and 34% of the total cultivated area whereby 11.6% of the total local population and 32% of the total labor force are directly engaged in the rice sector [5]. The rice sector contributes 7% to the national agricultural GDP [6], and therefore is associated with food and nutrition security and is critical for generating livelihood income for the Sri Lankan population.

Salinization of arable soils is an increasing challenge in global agriculture. Worldwide about 20% of total cultivated and about 33% of irrigated agricultural lands are afflicted due to soil salinity [7, 8]. Soil salinization is aggravated by the adverse effects of climate change causing abandonment globally of 0.3–1.5 million hectare $year^{-1}$ [9]. Consequently, an estimated 50% of the arable land would be salinized by the year 2050 [10]. The cost of global crop loss owing to salinization is estimated to be USD 27.3 billion [11]. Rice, with a threshold of 3 dSm^{-1} for most cultivated varieties, is very sensitive to salinity [12], and rice cultivation is extremely vulnerable to soil salinization. In moderately saline areas, the yield loss of rice is 10–15%, whereas in highly saline areas, the yield was reduced by 30–45% [13]. Soil salinity has become a major limiting factor for local rice production, especially in the irrigated farming systems in the north-central and eastern plains, and in the rain-fed systems in the west coast and the Jaffna peninsula [14]. The net income of rice production is reduced up to 22% and 43% in moderate and highly saline areas, respectively in Sri Lanka [15], and therefore salinity has a significant impact on the economic sustainability of local rice production systems.

1.2 The concept of soil salinity

Saline soils have an excess accumulation of Na^+, K^+, Ca^{2+}, Mg^{2+}, HCO_3^-, Cl^-, NO_3^-, SO_4^{-2}, and CO_3^{-2} or mixtures of these ions [16], thereby affecting the normal functions of plant growth. Based on the physiochemical properties: electrical conductivity (EC), exchangeable sodium percentage (ESP), sodium adsorption ratio (SAR), and pH, saline soils can be categorized as saline, sodic (alkali), and saline-sodic soils. Optimal soil conditions for crop growth are EC <4 dSm^{-1}, pH = 6.5–7.0, SAR <13, and ESP < 13 [17]. Saline soils are characterized by EC of saturated extract >4 dSm^{-1}, pH <8.5, SAR <13%, and ESP <15% of the exchangeable cations. However, saline soils often are in normal physical conditions with good structure and permeability, and therefore with proper management measures can be used for crop cultivation. In contrast, sodic soils are low in total salts but are characterized by high exchangeable Na^+. Sodic soils record high ESP >15%, pH > 8.5, and SAR >13%, but EC is often <4 dSm^{-1} [18, 19]. High levels of sodium and low total salts result in dispersed soil particles, and poor physical properties and sodic soils are sticky when wet, but hard, cloddy, crusty, and nearly impermeable to water when dry. Saline-sodic soils contain large amounts of total soluble salts but greater than 15% of exchangeable Na^+. The pH is <8.5, SAR >15%, and the EC is >4 dSm^{-1}. The physical properties of saline-sodic soils are good as long as an excess of soluble salts is present.

Soil salinity can arise due to natural processes causing primary salinity or due to manmade secondary salinity. In field conditions, low salinity *viz.* EC 2–4 dSm^{-1} can arise from natural salinity and/or irrigation salinity. Species with low to moderate salt tolerance can be grown successfully in low saline soils. Moderate to high salt-tolerant plant species can be grown when the EC is between 4 and 8 dSm^{-1} [20] usually present in water logged irrigated conditions. Under high salinity with an EC value of >9 dSm^{-1} only halophytes can be grown [21], and therefore the choice of crops is a useful measure of salinity management. Rice expresses low tolerance and based on current guidelines rice yields decrease by 12% for every unit of (dSm^{-1}) salinity increase above 3.0 dSm^{-1} [22, 23]. Detailed analysis of the salinity problem including associated intricate mechanisms, therefore, can help in developing effective management practices including the selection of tolerant rice varieties.

1.3 Major factors underlying salinity tolerance

Salinity tolerance in plants involves a number of traits that act in isolation (independent) or in combination (less independent) [24, 25]. The adverse effects of salinity on plant growth are generally associated with the osmotic potential of the soil solution and the high level of toxicity of sodium (and chloride for some species) that causes multiple disturbances in crop metabolism, growth, and development at the molecular, biochemical and physiological levels [26]. Plant response to salinity is expressed in two major phases: the initial, rapid osmotic phase that inhibits growth and, a later, slower ionic phase that accelerates tissue and organ senescence [27]. Salinity-induced osmotic effects reduce plant biomass and yields; however, selected ions, such as Na^+, Cl^-, Ca^{+2}, and Ba^+, cause additional injury and crop damage [28]. Accordingly, three distinct salinity response mechanisms were described in tolerant germplasm, including osmotic stress tolerance, Na^+ exclusion from photosynthetic and other sensitive tissues, and tissue tolerance against accumulated Na^+ and possibly also accumulated Cl^- [27].

Rice is relatively tolerant to salinity at germination and late vegetative growth, compared to the early seedling stage (3-leaf stage) and reproductive stage (pollination and fertilization). Therefore, apparently, plants at different developmental stages may express one or more of the different tolerance mechanisms. However, there is a poor correlation between the tolerance mechanisms expressed at the two most salt-sensitive stages, the early seedling and reproductive stages [27]. Multiple tolerance phenotypes, associated mechanisms, and genes were identified from different salinity-tolerant germplasm at different development stages [3, 29]. Therefore, a thorough understanding of the molecular mechanisms associated with tolerance traits, and advanced technological innovations to incorporate the traits into elite varieties can accelerate breeding programs targeting saline tolerance.

This chapter aims to review the literature available on the nature and the scale of soil salinization problems in local rice farming environments. The analysis, thereby, would provide actionable insights and potential targets to reduce the genetic vulnerability of elite germplasm and to improve soil salinity management strategies in local rice farming and production environments.

2. Rice farming systems in Sri Lanka

Rice is cultivated island-wide except at elevations 2000 m above sea level, and 75% of the rice lands in Sri Lanka are located in inland valley systems. The remaining 25% are in the terraced slopes of uplands, coastal plains, associated floodplains, and alluvial plains [30]. Rice cultivation is practiced following the "northeast" and "southwest" monsoons. The northeast monsoon is from December to February in the following year, and the southwest is from May to August. The spatial distribution of rainfall brought by the monsoons defines two distinct climatic zones namely the "wet" and "dry" zones. The wet zone is in the southwestern sector of the island, whereas the dry zone is in the north, northeastern, and eastern parts of the country. The marginal areas between the two major zones are in the intermediate zone (**Figure 1**). The wet zone receives rain from the southwest monsoon and the amount varies from about 100 mm to over 3000 mm. In addition, from March to April wet zone receives the "first inter-monsoon" rain due to convectional influence, which can be over 250 mm to >700 mm. The dry zone receives rain from the northeast monsoon and the peak rainy season is from December to February, where the rainfall can be up to 177 mm to 1281 mm. The influence of depression and cyclone weather systems in the Bay of Bengal cause a "second inter-monsoon" starting from late September [33] or October to November [33–35]. The entire island receives in excess of 400 mm of rain from the second inter-monsoon. The two inter-monsoon periods contribute 37% of the total annual rainfall [36] and are critically important for the local crop calendars.

The special variability in rainfall along with soil, elevation, and hydrological regimes identify diverse rice farming environments in the country. However,

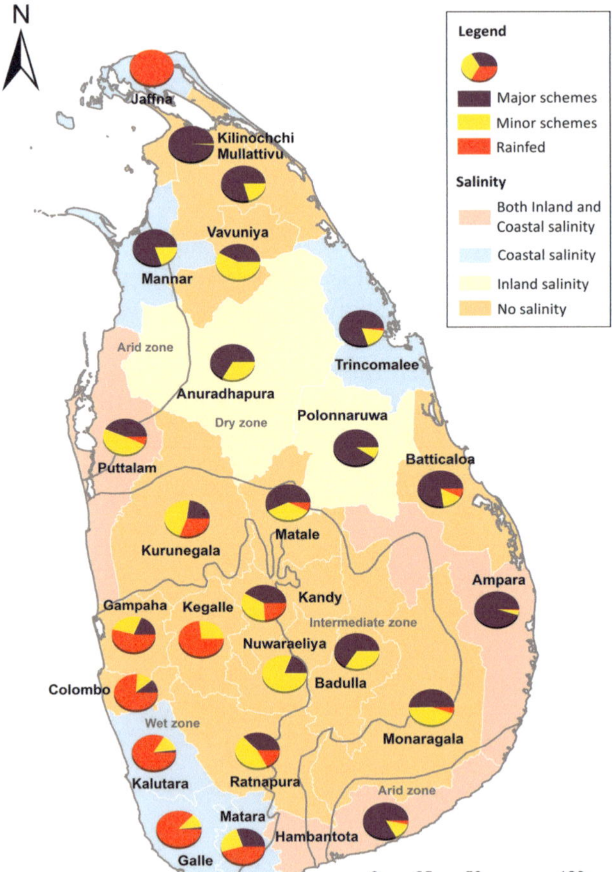

Figure 1.
Land extends of the different farming systems: Major irrigated, minor irrigated, and rain-fed based on data from 2011 to 2021. Lands affected due to coastal and inland salinity and the border lines of wet, dry, and intermediate zones are indicated (data obtained from [31, 32]).

hydrological regimes are the principal determinant of rice production in the country, and three major rice farming systems are identified based on the source, supply, and use of surface water. These include rice lands under major and medium irrigation schemes, minor irrigation schemes, and rain-fed systems (**Figure 1**). Major irrigation systems are those that have a command area of >1000 ha, whereas the command area of medium schemes is between <80 ha and 1000 ha. Small tanks or minor irrigation systems are those having an irrigated command area of 80 ha or less [37]. The different rice production systems are geographically distributed (**Figure 1**). Rice land irrigated by major and medium irrigation schemes extends in the dry zone in an area of 730,000 ha in the north-central and the eastern parts of the island [38], where the water supply is from major tanks, rivers, and major stream diversion systems. Rice farming in major irrigation systems contributes 53% of the total annual rice production. Rice farming systems in the central dry and intermediate zones mostly depend on minor irrigation. Minor irrigation schemes operate under village tanks that are natural or man-made water reservoirs. Out of the total extent of the irrigated lands, 37% is in this category, and the minor irrigated systems contribute 20% to the total annual rice production in the country [39]. In the southwest, southern coastal region of the wet zone, and in the Jaffna peninsula, rice is cultivated under rain-fed systems. Of the total area of rice lands, 34.7% is rain-fed [31], and rain-fed systems contribute only 27% to the total rice production. All three systems, therefore, are critical in providing food supply and livelihoods for the local community.

Rice cultivation in irrigated and rain-fed systems is practiced in two major cropping seasons namely "major cultivation season" and "minor cultivation season," coinciding with the two monsoons (**Figure 2**). The "major cultivation season", which is from September to February coincides with the second inter-monsoon rain followed by the northeast monsoon. In the major season, cultivation starts in late September [33] with the onset of the second inter-monsoon, and the crop is harvested at the end of February. In the minor cultivation season cultivation starts during the end of March to mid-April with the first inter-monsoon rain, and the crop is harvested at the end of August [33]. In the major cultivation season with sufficient rainfall, rice is cultivated throughout the country. However, during the minor cultivation season, 21% of the rice lands are abandoned or used for the cultivation of other field crops due to water scarcity [31]. Rice lands under major irrigated farming systems are cultivated in two seasons, *viz*. major cultivation season and minor cultivation season per year. However, in minor irrigation systems, rice is cultivated in one season, the major cultivation season, with a few exceptions where irrigation is sufficient to cultivate two seasons. Wet zone rain-fed systems in the southwest are cultivated in two seasons depending on water availability, but rainfed systems in Jaffna peninsula are cultivated only during the major cultivation season [31].

Average annual rice production from major cultivation season, major irrigation schemes during the period 2011 to 2022, was 1,592,588 MT accounting for 54% of the total annual rice production. In the minor irrigation and rain-fed rice farming systems, the average rice production from major season over the same period was 702, 953 MT (31%) and 653, 288 MT (29%), respectively [31]. Rice production was significantly reduced in the minor cultivation season due to the limitation of water in the minor irrigation and rain-fed systems whereby the average production from the minor season during the period from 2011 to 2022 was 340,634 MT and 136,014 MT, respectively [40]. Therefore, 72% of the total production from the minor cultivation season is from rice under major irrigation schemes, and the average annual production over the period from 2011 to 2022 was 1,222,106 MT [31]. Accordingly, altogether,

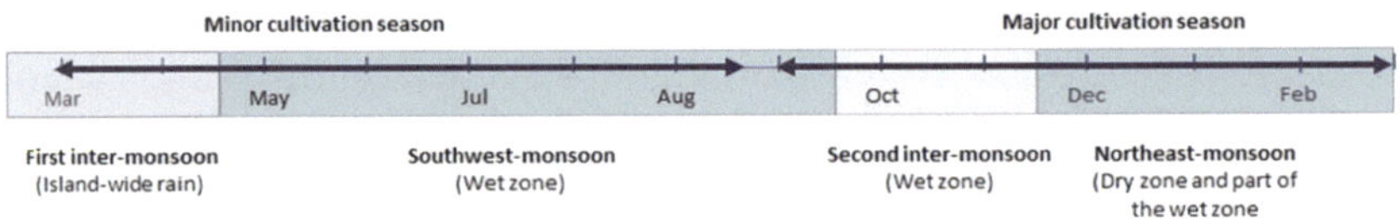

Figure 2.
Rainfall patterns and cultivation seasons in rice farming systems in Sri Lanka (adapted from [33–35]).

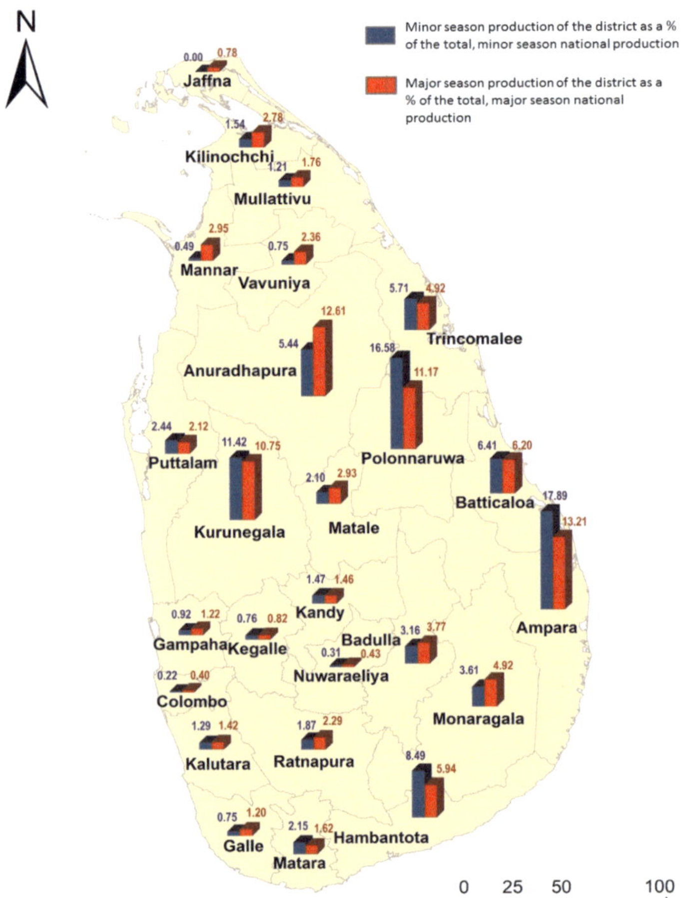

Figure 3.
Percentage of rice production in different districts in the two cropping seasons. Bar graphs indicate total seasonal rice production from the district as a percentage of the total national production from each season based on data from 2011 to 2021 (data obtained from [31]).

>70% of the national annual rice production is from major irrigated systems (**Figure 3**). Therefore, major season cultivation in the major irrigated fields is the most productive rice farming environment or the "mega-rice cultivation system" in the country. Although the national contributions are less from rain-fed and minor irrigated systems where smallholder farmers cultivate rice on a semi-subsistence scale, those systems are critically important for national food security.

3. Spatial variation of soil salinity in Sri Lanka

Soil salinity varies across the country and the highest salinity levels are reported in the north-central plains and along the coastal belt (**Figure 1**, **Table 1**) [15, 41–53]. Primary salinity due to natural weathering of salt-containing bedrock is restricted to isolated patches in the island. However, secondary salinity is widespread and found in inland and coastal rice farming systems. There are three main processes associated with secondary salinity in local rice farming systems *viz,* irrigation salinity, dryland salinity, and coastal salinity. Irrigation salinity arises in irrigated areas as a result of rising groundwater tables from excessive irrigation or due to the use of poor-quality water. In dryland non-irrigated landscapes, dryland salinity occurs due to the rise of the water table that drives soluble ions deposited by primary salinity to the soil surface. Coastal salinity is a result of the accumulation of salts in soil and water due to seawater intrusion from surface flow or seepage of seawater through sea level rise and tidal activity. The incidence of tides and dikes is the main reason for the intrusion of seawater into inland areas. Primary salinity, found in a few isolated sites such as the Nawagathegama soils series in the northwest part of the country [54], has a minimal economic impact on local agriculture; however, secondary salinity resulting from human activities causes significant economic losses in the local rice farming systems.

The inland salinity is a combination of both irrigation salinity and dryland salinity and has a mosaic distribution island-wide [53]. The main processes that contribute to irrigation salinity are the rise of groundwater tables resulting from poor drainage, excess irrigation, and perennial irrigation systems, which result in shallow water tables that bring salts to the upper layers of the soil profile. The problem is aggravated due to blocked drainage canals that prevent draining of salts [55]. A case study, in the largest irrigation scheme in the country the "Mahaweli," observed that soil salinity is a principal determinant of rice production in affected irrigated areas whereby yield loss of 10–15% is recorded in moderately saline areas; however, the yield was reduced by third in high saline areas [15]. In the segment "Mahaweli H" > 10% of the irrigable area is affected due to high salinity and an additional 40% is affected by moderate salinity providing a relative measure of the scale of the problem [15]. When salinity is >5dSm^{-1}, rice yield is reduced by more than 20% in affected compared to unaffected cultivations [56]. Inland salinity, therefore, has a significant impact on mega-rice cultivation environments, causing reduced production and direct economic losses.

Multiple studies have shown high concentrations of inorganic ions and EC >2.5 dSm^{-1} in irrigated water in the dry zone, significantly exceeding the recommended EC range in irrigation water, which is 0.7–0.75 dSm^{-1} [18]. Despite the high EC values, natural Na^+ concentrations in inland soils, including soils in the mega-cultivation environments, and except in a few isolated pockets, were low [55]. A significant portion of the northern half of the country recorded soil Na^+ concentration > 200 mg/kg, which is below the toxic limits [57]. However, high CO_3^{-2}, HCO_3^-, Ca^{+2}, and Mg^{+2} concentrations in the dry zone soils increase EC [58, 59]. Anuradhapura, Puttalam, Polonnaruwa, and Kurunegala districts record soil Mg^{2+} and Ca^{2+} concentrations as high as 730 ppm (optimum level for rice production, 120 ppm, [60] and 509–3452 ppm [61], respectively). Ca^{2+} saturation of cation exchange capacity (CEC) of soils in the dry zone is Ca 52.5% of CEC (5 cmolc/kg), whereas the recommended Ca CEC is <20% [62]. Further, high HCO_3^{-2} and SO_4^{-2} are reported in water in irrigation wells in north-central areas, which can contribute to high soil EC values. In such conditions, although not subjected to Na^+ toxicity, the crops are subjected to osmotic stress owing to the high ion concentration in the root zone.

Type of salinity	Location	Salinity level (Electrical conductivity)	Temporal variation	Other significant findings	Reference/s
Coastal salinity	Jaffna Peninsula	Soil EC range: 0.05–34 dSm^{-1}	August 2019 (Peak of the dry season and fallow season of the paddy cultivation)	13% and 46% of the paddy cultivation lands are extremely saline and saline respectively, Salinity increased by 1.4 folds in 30 years Salinity increment- Slightly saline area by 51% Moderately saline area by 24% Extremely saline area by 65%	[41–43]
	Jaffna	EC of soils collected from Tsunami affected areas high (7–19 dSm^{-1}) in January and low (1–5 dSm^{-1}) in September	High salinity in January and low in September	EC of water and soil was high just after Tsunami and decreased to low levels with time.	[44]
	Bentota river basin Bentota DSD- the lower part of the river basin	Soil EC range- 0.14–20 dSm^{-1} 0.0112 to 0.2443μScm^{-1}	August 2016 to March 2017 Groundwater salinity is high in September and from January-March	Land in the Bentota River basin Non-saline- 29.79% Slightly saline- 40.58% Moderately saline- 19.47% Saline- 10.16% The EC distribution pattern was changed with the depth of the soil layer. 40 cm and 60 cm reported high salinity levels Land in Bentota DSD Moderately saline- 35.6% Slightly saline- 64.4%	[45, 46]
	Hambantota district	Soil EC range > 16 to 0 dSm^{-1}	October-December (Dry period) only	Non- saline- 40.4% (<2) Slightly saline- 38.0% (2–4) Moderately saline- 20.5% (4–8) Strongly saline- 1.2% (8–16) 24.8 sq.km (>16)	[47]
	Colombo Kahapola	Soil EC (1:5) 0–2.8 dSm^{-1}	Higher salinity >2 dSm^{-1} in the dry season August-September	EC values greater than 0.6 dSm^{-1} in 1:5 Extract.	[48]
	Puttalam-Madampe	Soil EC 3–4 dSm^{-1}	Highest salinity August-September		
	Mannar	Soil EC 3–4 dSm^{-1}	—		

Type of salinity	Location	Salinity level (Electrical conductivity)	Temporal variation	Other significant findings	Reference/s
	Mathara	Soil EC 100–422 μscm^{-1}	Highest salinity August-September		
	Mathara (Tsunami affected and unaffected areas)	Soil EC in the tsunami-affected area varied from 100 μScm^{-1} to 422 μScm^{-1} and groundwater salinity changed between 400 μScm^{-1} to 2000 μScm^{-1}	—	Soil depth soil salinity, pH, and EC variations were directly related to groundwater salinity	[49]
	Puttalum district-Arachchikattuwa	Maximum E C values 1.9, 1.8 and 1.7 dSm^{-1}	Maximum EC- February, July, and August Minimum EC- April, September, October, and November	Percentage land extent EC >4 dS/m was 10% in September and April. April (secondary cultivation season) and September (Main cultivation season) were the suitable months to establish paddy crop since EC levels were at a minimum	[50]
	Nilwala River area	Groundwater EC 38.48 μScm^{-1}	—	Due to sand mining	[51]
Inland salinity	Kurunegala District Mahananeriya Ibbagamuwa Alawwa	Major part of the areas was < 0.15 dSm^{-1} In certain areas EC (1:5) rise up to 0.8 dSm^{-1}	—	Soil chemical parameters in the regions were suitable for paddy cultivation.	[52]
	Mahaweli river system H Irrigation scheme Nochchiyagama Madatugama	Soil EC range- 1.3–7.90 dSm^{-1} Soil EC range- 1.1–8.80 dSm^{-1}	September to February March to August (The main inflow seasons from Mahaweli)	Mahaweli H area has less than 10% of the total irrigable area with significant soil salinity problems 40% of the farmer fields are affected by moderate salinity The yield loss ranged from 10–15%; high and severe salinity reduced yield by about one third	[15]

Table 1.
Spatial and temporal variation in salinity in Sri Lanka.

Coastal salinity has a significant impact on the rain-fed farming systems in the southwest coastal belt [48, 63] and in the Jaffna peninsula [41–43]. Due to the low coastal slope rice farming systems in the southwest coast are frequently affected by inundation and irrigation with salty water [48, 52, 64]. An area of 0.112 million ha is affected due to coastal salinity in Sri Lanka [30], where the EC of the soil extraction can exceed 4 dSm^{-1} [65]. In affected sites, tidal waves can range between 45 cm and 60 cm during spring tide and between 10 cm and 25 cm during neap. In dry conditions, the reduction of water levels in rivers causes backflow of seawater along rivers and intrudes the rice fields with salt water, thereby saltwater intrudes into lands up to 50 cm above mean sea level [32] in dry weather. Further, blockage of drainage canals can also cause salinization in rice fields in the coastal line. The EC range in the southwest coast varied from 0.1 to 0.4 dsm^{-1} based on soil: water 1:5 extracts. Unlike in inland salinity, soils affected due to coastal salinity contain high concentrations of Na^{+} and Cl^{-} ions [66, 67]. In the southwest coastal belt (Kaluthara to Matara), Na^{+} levels vary from 50 to 100 ppm; however, in Jaffna and on the eastern coastline the Na^{+} levels are higher than 200 ppm [53]. Detailed observations were made in the Bentota river basin, where 70% of the land is affected due to coastal salinity, ranging from 8 dSm^{-1} to 16 dSm^{-1} along with >10% of land affected by high salinity (>16 dSm^{-1}) [45, 46, 53]. Agricultural production loss per year due to salinity in the area is over 3.6 million USD [45]. Rice lands are gradually abandoned in the Jaffna peninsula owing to the high levels of seawater intrusion causing salinization of soils and water in wells used for irrigation. In addition, the major irrigated farming systems in Mannar, Hambantota, and Trincomalee coastal lines are also affected due to coastal salinity, where EC values can vary between 3 to 30 dSm^{-1} [42, 47, 48, 68]. Coastal salinity has a significant impact on semi-subsistence rice farming systems in the southwest, northwest, and eastern coastal line limiting food availability and livelihoods for the local farmers, and thereby destabilizing local food systems and increase poverty.

4. Temporal variation of salinity in Sri Lanka

Soil salinity varies seasonally whereby both coastal and inland salinity is high during the dry season. During the wet season with high rainfall salts are washed off from the soils, but in dry periods, salts accumulate causing salinity. Furthermore, the impact of coastal salinity is reduced during the wet months due to high rainfall causing a pushing effect and runoff inflow toward the sea reducing sea water backflow and saline water intrusion into the land [69]. With the heavy rain, salts are washed off in the main cultivation season, but with less water availability and with high water evaporation, the minor cultivation season is more salinity prone especially toward the end of the season. The phenomena create a seasonal cycle in salinity in the irrigated and rain-fed systems whereby salinity deposited from primary and secondary sources (seawater and irrigation water) during the dry season is washed off during the wet periods (**Figures 4** and **5**) [48, 50, 69–71]. The amount of rainfall and water availability, therefore, are primary factors that determine soil salt concentration and the variability in rainfall causes seasonal and annual fluctuations in salinity levels.

The west coast where the rainy (March to August: first inter monsoon followed by southwest monsoon; September to February: second inter-monsoon followed by northeast monsoon) and dry periods observe bimodal patterns. Two clear salinity peaks were recognized. In time series salinity estimates of soil and water in selected sites in the rain-fed and irrigated rice lands in the west coast the highest salinity levels are recorded from August to September at the end of the minor cultivation season (EC > 24 dSm^{-1}). A second minor salinity peak was observed at the end of the major

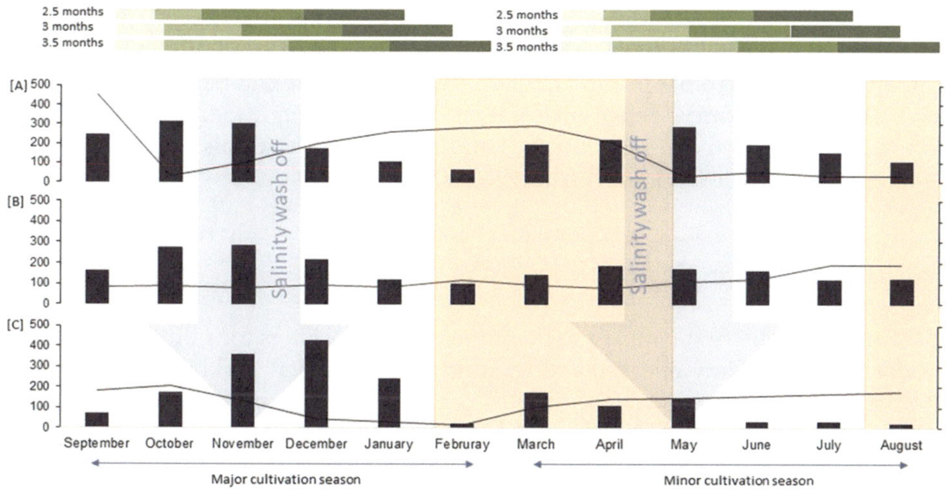

Figure 4.
Temporal variation in salinity and rainfall in Northwestern [A], Southwestern [B], and eastern [C] coastal line of Sri Lanka. The vertical axis on the left indicates the monthly average rainfall (mm) and on the right indicates soil electrical conductivity (dSm^{-1}). Gray boxes indicate monthly rainfall (mm), whereas the line electrical conductivity (dSm^{-1}). The green boxes from left to right present crop stages of 2.5, 3-, and 3.5-months rice varieties, including seedling (susceptible), tillering/vegetative, reproductive/PI, and mature/ripening crop. The crop-salinity susceptibility window is indicated in red boxes (based on [45, 48, 50, 63]).

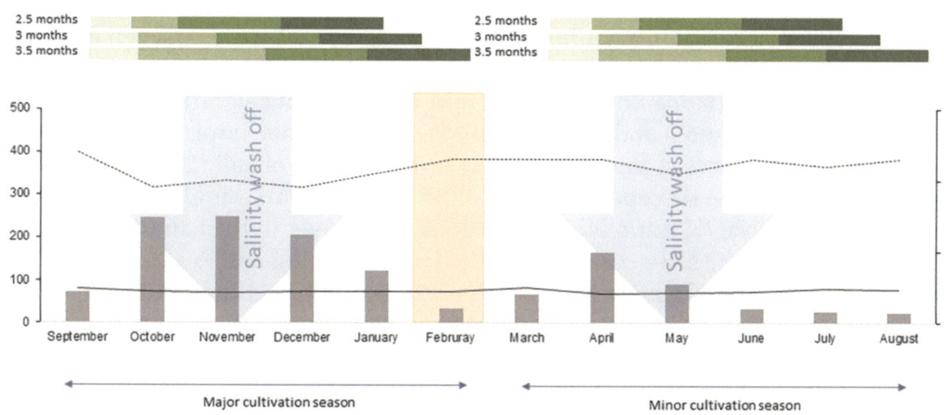

Figure 5.
Temporal variation of salinity and rainfall in irrigated "mega production areas" of Sri Lanka. The vertical axis on the left indicates monthly average rainfall (mm) and on the right indicates the electrical conductivity (dSm^{-1}) in irrigation water. Gray boxes indicate the rainfall, whereas the line indicates electrical conductivity (dSm^{-1}) in irrigated water. The green boxes from left to right present growth stages of crops of 2.5, 3-, and 3.5-months, including seedling (susceptible), tillering/vegetative, reproductive and PI (susceptible), and mature/ripening stages. The crop/salinity susceptibility window is indicated in red boxes (based on [69–71]).

cultivation season during February to March, where the EC values in soil >20 dSm^{-1} (**Figure 4**) [45, 48, 50, 63]. A similar pattern is reported in the Jaffna peninsula [72]. When salts were not washed off due to insufficient rain, the seedling crop of the both minor and major cultivation season is affected due to early season salinity. Unlike in the west coast, in the north-central and east parts of the country EC values gradually built over a period of six months during the dry season starting from the later part of the major cultivation season throughout the minor cultivation season [72] (**Figure 5**). Accordingly, the major irrigated rice farming systems are affected due to high salinity over the dry months from April to October, where EC value can be as high as 7 to 9 dSm^{-1}. Therefore, farmers experience transitional saline conditions with significant yearly (depending on the annual rainfall) and seasonal fluctuations in salinity level in both irrigated and rain-fed local farming systems.

5. Salt-tolerant germplasm in Sri Lanka

Sri Lanka was the first country to recommend rice varieties for salinity-prone areas whereby the first variety released was an improved land race "Pokkali" [73]. Presently, there are four elite locally improved rice varieties recommended for cultivation in salt-affected areas namely Bg369, Bg310, At401, and At354 [33]. Salt- tolerant germplasm Pokkali [74] and Nona Bokra [75] are extensively used in local salinity tolerance breeding programs. Pokkali is an immediate parent of Bg310, At401, and At354, whereas Nona Bokra is an immediate parent of Bg369. In addition, tolerant and moderately tolerant local rice germplasm that express Na^+ exclusion at root surface, Na^+ exclusion at root xylem parenchyma, and capacity to maintain shoot water status under saline stress were identified [76].

Pokkali and Nona Bokra are known to withstand salinity EC > 12 dSm^{-1} from seedling to harvesting [68, 77–79] and were also reported to express high to moderate salinity tolerance throughout the crop life. Pokkali and Nona Bokra seedlings recorded salinity susceptibility scores of three compared to a score of seven in susceptible checks [80]. The concentration of NaCl inhibiting 50% of shoot growth (IC50) in 72 hours is 305 mM in Nona Bokra and 272 mM in Pokkali but significantly low in susceptible IR64 (188 mM) and IR29 (243 mM). Shoot Na^+ accumulation after 2 hours of 200 mM salinity stress at the seedling stage was 50% in Pokkali (2629.6 ppm) compared to that in the susceptible variety IR64 (5674.3 ppm). Similarly, plant Na^+ concentration and Na^+/K^+ ratio were 2.4 and 2.5 folds lower in Pokkali seedlings compared to that in the salt-sensitive, IR64 [81]. Nona Bokra and Pokkali report low shoot Na^+ content of 5 mg/g by shoot dry weight compared to IR64 (13 mg/g) and IR29 (12 mg/g) [82]. Nona Bokra, Pokkali, and IR28 reported 0.6 mg, 4.0 mg, and 5.8 mg reduction of dry weight of a fertile spikelet at 15 dSm^{-1}, indicating the presence of tolerance mechanisms that are effective over the growth period. [83]

Pokkali and Nona Bokra express "Na^+ exclusion," and therefore the phenotypes are "high Na^+ in the culm and low Na^+ in the leaf" [76]. Evidence found on additional resistance traits such as efficient photosynthesis, adaptive defense mechanisms, osmolyte accumulation, membrane integrity, and optimum biochemical environment in the cells of Pokkali may provide whole-life resistance [84–87]. Furthermore, multiple Quantitative Trait Loci (QTLs) associated with salinity tolerance were predicted in Pokkali [88–92], including the *Saltol* locus [93–97], which is collocated with a major QTL, *SKC1* from the salt-tolerant landrace, Nona Bokra [98]. *Saltol*/*SKC1* is a major QTL associated with low Na^+ uptake, high K^+ uptake, and Na^+/K^+ homeostasis in shoots in the seedling stage. *Saltol/SKC1* was collocated with an important candidate gene *OsHKT1;5*, a high-affinity K transporter gene [99–101] that is expressed in xylem parenchyma cells, and retrieves Na^+ ions from xylem sap under salinity. Furthermore, root QTLs associated with the total Na^+ in the root (*qRNTQ-1*) and K^+ concentration in root (*qRKC-4*) that contribute to salt tolerance were reported in a cross between Nona Bokra/Koshihikari [102]. The physiological, genetic, and genomic studies, therefore, revealed multiple tolerance mechanisms expressed at different growth stages in Pokkali, Nona Bokra, and in the local varieties derived from those parents.

Although the four recommended local varieties expressed comparable saline tolerance at specific crop growth stages, overall tolerance levels during the crop life were not comparable with that of Pokkali and Nona Bokra. Based on the percentage germination of seeds soaked at 45 dSm^{-1} salt solution for 9 days [103], At 354 (21–30% germination) was moderately susceptible for salinity compared to Pokkali (51–70% germination); however, Bg369 was highly susceptible (>10% germination) [68, 78, 104]. Evidence from multiple studies found that based on susceptibility scores at 12 dSm^{-1} At354 and Bg369 were tolerant and comparable with Pokkali at seedling stage [68, 76]. Nona Bokra and Pokkali seedlings survived (100% survival rate) up to 8 dSm^{-1}; however, above 4 dSm^{-1} survival rates declined by 50% in IR28 [105]. Based on the survival percentage of plants treated at 12 dSm^{-1} from 3 weeks to flowering, Bg369 and At354 were comparable with Pokkali [68]. Additionally, the yield reduction at 8 dSm^{-1} based on grain yield per panicle, At354 was low compared to Pokkali and Nona Bokra (per panicle grain yield reduction was 0.72 g, 0.43 g, and 0.38 g respectively) [78]. A F5, RILs population of At354 × Bg352 (susceptible check), predicted six QTLs on chromosomes 1 and 4 from At 354, contributing to 10–16% of the total phenotypic variability. The six QTLS were collocated with 10 tolerant haplotypes of Os01g0581400, Os10g0107000, Os11g0655900, Os12g0622500, and Os12g0624200, highlighting the genetic potential of At354 as a salinity tolerant variety [106]. Therefore, it is clear that the full potential of Pokkali, Nona Bokra, and the related germplasm is yet to be explored and incorporated into local elite rice varieties.

Several reports have identified unique local rice germplasm that can be used in base broadening for salinity tolerance [68, 78, 104]. In a study, using 102 rice accessions from local germplasm including both traditional and improved varieties five susceptibility/tolerance categories were identified based on growth parameters during the initial phase (Phase I) of osmotic stress caused by reduced water uptake due to excess salts in the external soil solution at 100 mM Na^+(**Figure 6**) [65]. Interestingly, in this study Pokkali was categorized as "tolerant," whereas six phenotypically superior germplasm were categorized in the "highly tolerant" category. Multiple tolerance traits were expressed in 7% of the germplasm [76]. Interestingly, unique germplasm

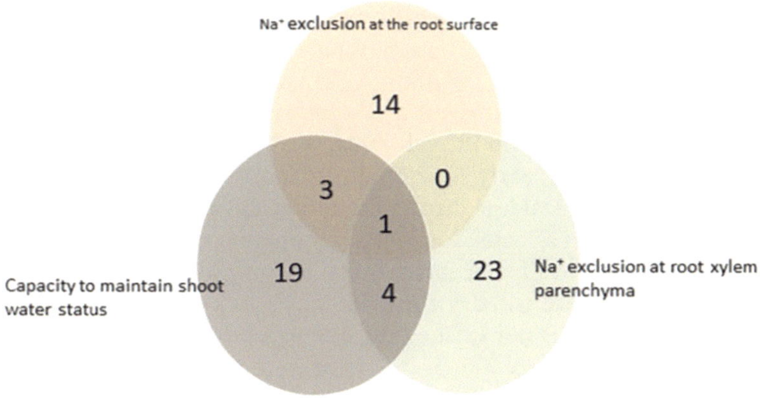

Figure 6.
Salinity tolerance mechanisms in 54 selected traditional and improved rice germplasms from Sri Lanka at the seedling stage (based on [76]).

superior to Pokkali at different development stages were also reported. Bg406 records 70% germination under 12 dSm^{-1} EC, expressing similar or more tolerance than Pokkali (51–70%); At402, performed equal or better at the seedling stage and reproductive stages [68]. Similarly, At303, Bg350, and Bg450 have a high tolerance compared to Pokkali at the osmotic phase of salinity stress [76] owing to their ability to maintain high relative growth rate at early stages of salinity (24 hours after final salinization at 10 dSm-1). The unique varieties namely Bw400, At402, and Bg406 expressed tolerance compatible with Pokkali at varying degrees throughout the crop life [107]. As these traits and associated alleles are distinct from Pokkali and Nona Bokra parental germplasm, these provide unique germplasm base for increasing salinity tolerance in elite rice genotypes without compromising crop vigor. Therefore, further analysis of the local germplasm will identify novel physiological and genetic tolerance mechanisms, and the associated genotypes can be used in base-broadening for salinity tolerance breeding in rice.

6. Discussion

Secondary salinization is increasingly becoming a productivity constrain in local rice farming systems in the “mega-cultivation environments” in north central and eastern inland fields and rain-fed systems in the west, southwest and eastern coastal line. The spatial and temporal dynamics in climatic, edaphic, and hydrological regimes, and associated variations in the farming practices create complex and diverse saline stress environments in local rice farming systems. Furthermore, salinity stress coincides with the susceptible 20-day period, the 3-leaf stage and PI of the rice crop life whereby early season salinity coincides with the 3-leaf seedling stage and late season salinity coincides with the PI stage. Secondary salinization of cultivation environments, therefore, is the result of the interaction among multiple factors, including sea intrusion, sub-optimal quality of the irrigation water, injudicious use of irrigation water, suboptimal farming practices, and unavailability of local elite germplasm, and hence is a problem that operates at system level; therefore, system-level interventions are necessary to mitigate and/or to manage the impacts.

Analyzing the variability in physical and chemical field characteristics and crop growth stage enabled zoning of the saline-afflicted production environments into three major categories: (1) late-season salinity in irrigated mega-cultivation environments during the minor cultivation season, where the soil EC values can be higher in selected locations but on average $\geq 7\ dSm^{-1}$, (2) late season salinity in rain-fed farming systems in the west, southwest, and eastern coastal line during the minor cultivation season, where the soil EC values can be higher in selected locations but on average $\geq 20\ dSm^{-1}$, and (3) early season salinity in rain-fed farming systems in west, southwest. and eastern coastal line during the minor cultivation season, where the soil EC values can be higher in selected locations but on average $\geq 12\ dSm^{-1}$. Although, detailed studies were not available, early season salinity during the major season is reported in the intermediate and dry zone rice fields as a result of residual salts from the previous season that was not washed off due to inadequate rainfall from north-west monsoon (personal observation).

Mega-cultivation environments that contribute nearly 70% of the average annual rice production [108] is an important economic enterprise and critical for the local economy and food availability. However, the intensive irrigated farming system is destabilized as a result of secondary salinization. Irrigation practices in the absence of proper drainage management trigger accumulation of salts in the root zone negatively affecting crop productivity. Poor drainage, undulating topography, reuse of irrigation water, and mono-cropping for long periods are recognized as the main causes of salinity developments in inland areas [109]. The seasonal soil salinization and desalinization cycles are driven by the deposition of salt by irrigation and evaporation and the removal of salts from the root zone by washing off due to rain and by gradual leaching. However, imbalances in this seasonal cycle can create residual effects leading to the accumulation of salts in the system, increasing residual salinity levels, and gradual salinization of the soils. As late-season salinity is largely an irrigation-limited issue, optimizing farm irrigation management strategies and robust water governance can reduce salinity impacts significantly and enhance productivity, while minimizing long-term land degradation in rice farming systems. Standards for irrigation and irrigation water quality based on rainfall, edaphic, and crop parameters [110–113] can be useful for efficient use of irrigation water and also sustainable land use. Actionable guidelines for irrigation and management of water quality are, therefore, an immediate need to prevent or minimize rapid soil salinization in mega-cultivation environments.

Furthermore, late season crop in mega-rice cultivation environments is subjected to simultaneous occurrence of multiple abiotic stresses, including water stress, high temperature, and salinity. Sterility of panicles is very common in saline-afflicted rice environments possibly as a consequence of a decreased pollen viability or receptivity of the stigmatic surface or both [114–117]. Salinity stress at the reproductive stage, especially after the booting stage, cause a reduction of grain yield due to the reduced number of filled grains, total number of grains and total grain weight per panicle, 1000-grain weight, and total grain weight per plant [118–120]. Evidence found that the reduction in grain number and grain weight in salinized panicles was not only due to reduced pollen viability but also due to higher accumulation of photosynthates (sugars) in panicle branches and panicle stem coupled with reduced activity of starch synthetase in developing grains. Accumulation of Na^+ in floral parts impedes starch synthase activity that transfers glucose to starch primer in developing grains resulting in the failure of the seed set [121]. However, early grain initiation, loss of pollen fertility, failure of pollination, spikelet death or zygotic abortion, changes in carbohydrate availability, and changes in the kernel sink potential are also the consequences of water deficit [122, 123]. The optimum temperature range for normal growth of rice is 22 to 32°C [124] and the threshold temperature of rice during anthesis is 33.7°C [125]. Under local open field conditions, pollen sterility was recorded at 31 ± 0.8°C [126]. Temperature more than 35°C under high relative humidity (near 85–90%) has an effect on the evaporative cooling of spikelet and makes complete spikelet sterility, and thus subsequent yield losses [127]. Therefore, designing mitigation measures would require a holistic approach whereby management practices and technology interventions need to be integrated effectively to avoid, overcome, or prevent the parallel occurrence of multiple abiotic stresses during the late season.

Early minor cultivation season, and occasionally in selected fields during major season, salinity can affect the seedling crops as a result of residual salinity from the previous season that is not washed off owing to insufficient rainfall or due to continuous seawater intrusion causing inundation of coastal low-lying rice lands. Early season salinity devastates seedling fields causing total loss of the crops. Transplanting aged seedlings is commonly practiced in coastal rice fields to avoid early-season salinity whereby farmers transplant 18 days old seedlings instead of 12 days and practice gap filling in case of mortality. Old seedlings with an established root and shoot system can tolerate salinity stress [128]. In addition, at manageable levels, salinity-afflicted coastal rice lands can produce rice by the cultivation of tolerant cultivars such as At354, At401, and Bw400 and by practicing recommended good agricultural practices (GAP) [113]. These include proper management of saltwater drainage canals, proper management of soil, maintenance of satisfactory fertility levels, pH and structure of soils by maximization of soil surface cover, application of organic manure, using crop rotation, minimum tillage, proper leveling, and adding gypsum ($CaSO_4$), which can cause leaching out of excess sodium [55].

Despite being one of the most susceptible cereal crops, rice cultivation was traditionally practiced in coastal floodplains as an effective way of land utilization in South and East Asia [129, 130], including Sri Lanka. In experimental plots, salinity of soil was controlled below 3% by growing rice. Soil desalination rate was increased from 65 to 74% by combining physical remediation measures with planting of rice [131]. The reason for this counterintuitive practice is that rice thrives well in standing water, which helps in leaching salts from the root zone to lower layers. Persistence of multiple abiotic stresses, including waterlogging in the wet season and soil and water salinity in both wet and dry seasons make managing these rain-fed production systems challenging. Therefore, salt-tolerant varieties, while increasing yields, enable effective use of saline lands. Integrated approaches based on a detailed understanding of the land potential can help in effective land diversification and crop diversification for developing sustainable land use strategies, and cultivation of tolerant rice varieties can be part of the solution.

Zoning of saline-afflicted production environments enables identifying genetic vulnerabilities, and thereby the unique targeted phenotypes (ideotypes) and locally adapted varieties. Ideotype breeding can create multi-trait genotypes to increase genetic yield potential by modifying selected individual traits to improve crop performance in saline production environments. Since salinity levels in local farming systems are transitional with significant yearly and seasonal fluctuations, maximizing resource use and profits in farms would require a clear understanding of the tradeoffs between performance of varieties under benign conditions and under stress. The vigor/ stress response tradeoff of the varieties, therefore, is an important selection criterion for varietal selections. Knowing the vigor/stress response tradeoff and the traits associated with the tradeoff enables designing potential ideotypes. Accordingly, three major ideotypes can be described. Two different ideotypes can target late-season salinity in irrigated mega-cultivation environments *viz.* (1) high yielding, 3 to 3.5 months varieties that are tolerant at >7 dSm-1, for intensive irrigated farming systems affected due to late season salinity (PI stage) and (2) high yielding, 2.5 to 3 months varieties for intensive irrigated farming that can avoid late season salinity. (3) the third, ideotypes targeting the rain-fed system in the west, with southwest, and the eastern coastal line will have to express tolerance throughout the crop life including seedling and PI stages. Varieties with average yield under salinity up to 12 to 20 dSm^{-1} would be a practical selection criterion for breeding varieties for the semi-subsistence rain-fed cultivation systems in these sites.

Climate change and rising temperatures have a strong positive correlation with increasing soil salinity [132]. Sri Lanka being a tropical island is highly vulnerable to climate change whereby the average temperature is increasing at a rate of 0.01–0.03°C per year [133–135] along with significant fluctuations in rainfall resulting in frequent droughts and floods [34, 136–138]. Higher temperature intensifies the excessive deposition of salt on the surface due to evaporation and increased capillary action. Once reached the surface leaching salts below the rooting zone is extremely difficult, especially under water-limited conditions in the dry zone. Moreover, as an indirect effect of increased temperature sea level rises resulting in increased salinity encroachment in coastal and deltaic areas that have previously been favorable for rice production [138]. The sea level rise in Sri Lanka is 0.3 m by 2010 with a predicted increase of sea level up to 1 m by 2070 [139]. Natural land degradation in coastal areas through sea level rise and increased coastal salinity has changed land use patterns [140]. In the western coast, in addition to rising sea level a combined impact of multiple factors, including groundwater extraction, river damming for hydropower, and riverbed mining has caused sinking of the shorelines and drawing seawater inland.

7. Conclusion and prospects

The temporal and spatial impact of salinity signifies a wide variation in different rice farming systems in Sri Lanka. The immediate and future risk assessments can scale the increasing trends in secondary salinization of the local rice farming systems that are further aggravated by climate change. Such actionable data can help in designing efficient management strategies.

Acknowledgements

This work was carried out with the aid of a grant from UNESCO and the International Development Research Centre, Ottawa, Canada. The views expressed herein do not necessarily represent those of UNESCO, IDRC or its Board of Governors.

Author details

Bhagya Indeevari Dasanayaka[1], Demuni Sumith de Zoysa Abeysiriwardena[2] and Chandima Kumudini Ariyarathna Hanchaplola Appuhamilage[1,3]*

1 Postgraduate Institute of Science, University of Peradeniya, Peradeniya, Sri Lanka

2 CIC Agribusiness Center, Pelwehera, Dambulla, Sri Lanka

3 Department of Botany, Faculty of Science, University of Peradeniya, Peradeniya, Sri Lanka

*Address all correspondence to: chandimaa@sci.pdn.ac.lk

References

[1] Khush GS. What it will take to feed 5.0 billion rice consumers in 2030. Plant Molecular Biology. 2005;**59**:1-6. DOI: 10.1007/s11103-005-2159-5

[2] Food and Agricultural Organization (FAO). Rice in global markets. In: Proceedings of the FAO Rice Conference; 12-13 February 2004. Rome, Italy: FAO; 2004. pp. 345-376

[3] Liu L, Han J, Xu X, Xu Z, Abeysinghe KS, Atapattu AJ, et al. Dietary exposure assessment of cadmium, arsenic, and lead in market rice from Sri Lanka. Environmental Science and Pollution Research. 2020;**27**: 42704-42712. DOI: 10.1007/s11356-020-10209-0

[4] Senanayake SMP, Premaratne SP. An analysis of the paddy/rice value chains in Sri Lanka. Asia-Pacific Journal of Rural Development. 2016;**26**(1):105-126

[5] Central Bank of Sri Lanka. Annual report 2010. 2011 [Internet]. Available from: https://www.cbsl.gov.lk/en/publications/economic-and-financial-reports/annual-reports/annual-report-2010

[6] Central Bank of Sri Lanka. Annual report. 2020. Available from: https://www.cbsl.gov.lk/en/publications/economic-and-financial-reports/annual-reports/annual-report-2020

[7] FAO. Tracking Progress on Food and Agriculture-Related SDG Indicators 2021: A Report on the Indicators under FAO Custodianship [Internet]. Rome, Italy: Food and Agriculture Organization of the United Nations (FAO); 2021. Available from: http://www. fao.org/sdg-progress-report/en

[8] Wang Z, Cheng J, Chen Z, Huang J, Bao Y, Wang J, et al. Identification of QTLs with main, epistatic and QTL× environment interaction effects for salt tolerance in rice seedlings under different salinity conditions. Theoretical and Applied Genetics. 2012;**125**:807-815. DOI: 10.1007/s00122-012-1873-z

[9] Harper RJ, Dell B, Ruprecht JK, Sochacki SJ, Smettem KRJ. Salinity and the reclamation of salinized lands. In: Stanturf JA, Callaham MA, editors. Soils and Landscape Restoration. Academic Press. 2021. p. 193-208

[10] Jamil A, Riaz S, Ashraf M, Foolad MR. Gene expression profiling of plants under salt stress. Critical Reviews in Plant Sciences. 2011;**30**(5):435-458. DOI: 10.1080/07352689.2011.605739

[11] UN FAO. Global Soil Partnership. 2021. Available from: https://www.fao.org/global-soil-partnership/en/

[12] Maas EV, Hoffman GJ. Crop salt tolerance—Current assessment. Journal of the Irrigation and drainage division. 1977;**103**(2):115-134. DOI: 10.1061/JRCEA4.0001137

[13] Sarangi SK, Maji B. Sustainable rice cultivation in coastal saline soils: A case study. Institute India. In: Achieving Sustainable Cultivation of Rice. Cambridge: Burleigh Dodds Science Publishing; 2017. p. 101-136. DOI: 10.19103/AS.2016.0003.07

[14] Abeysiriwardena DDZ. Rice varietal improvement for future challenges. In: Rice Congress. Peradeniya, Sri Lanka: Department of Agriculture;. 2000. p. 87-99

[15] Thiruchelvam S, Pathmarajah S. Economic analysis of salinity problems

in the Mahaweli River system H irrigation scheme in Sri Lanka. EEPSEA research report series/IDRC. Regional Office for Southeast and East Asia, Economy and Environment Program for Southeast Asia. 1999. Available from: https://EconPapers.repec.org/RePEc:eep:report:rr1999082

[16] Rath KM, Fierer N, Murphy DV, Rousk J. Linking bacterial community composition to soil salinity along environmental gradients. ISME Journal. 2019;**13**(3):836-846. DOI: 10.1038/s41396-018-0313-8

[17] Reitsma KD, Schumacher TE, Clay DE. Soil Quality. In: Deneke D, editor. Alternative Practices for Agronomic Nutrient and Pest Management in South Dakota. South Dakota, USA; 2011. pp. 1-8

[18] Richards LA. Diagnosis and improvement of saline and alkali soils. Soil Science Society of America Journal. 1954;**18**(3):348-348

[19] Davis JG, Waskom RM, Bauder TA, Cardon GE. Managing sodic soils. Colorado State University Cooperative Extension. Crop Series Soil, Fact sheets no. 0.504. 2007. pp. 1-5

[20] Rao A, Ahmad SD, Sabir SM, Awan SI, Shah AH, Abbas SR, et al. Potential antioxidant activities improve salt tolerance in ten varieties of wheat (Triticum aestivum L.). American Journal of Plant Sciences. 2013;**4**(6):69-76. DOI: 10.4236/ajps.2013.46A010

[21] Tilaki GAD, Gholami F, Bezdi KG, Behtari B. Germination percentage and recovery of Lolium perenne L. and Bromus tomentellus Boiss (Poaceae, Liliopsida) seeds at several osmotic potential levels of iso-osmotic solutions. Поволжский Экологический Журнал. 2014;**2**:284-292

[22] Maas EV, Grattan SR. Crop yields as affected by salinity. In: van Skaggs RW, Schilfgaarde J, editors. Agricultural Drainage. USA: American Society of Agronomy, Wiley; 1999. pp. 55-108. DOI:10.2134/agronmonogr38.c3

[23] Grattan S, Hanson B. Crop salt tolerance. In: Hanson B, Grattan SR, Fulton A, editors. Agricultural Salinity and Drainage. University of Califonia Irrigation Program. Davis: University of California; 1999. pp. 15-22

[24] Ismail AM, Heuer S, Thomson MJ, Wissuwa M. Genetic and genomic approaches to develop rice germplasm for problem soils. Plant Molecular Biology. 2007;**65**:547-570. DOI: 10.1007/s11103-007-9215-2

[25] Moradi F, Ismail AM. Responses of photosynthesis, chlorophyll fluorescence and ROS-scavenging systems to salt stress during seedling and reproductive stages in rice. Annals of Botany. 2007;**99**(6):1161-1173. DOI: 10.1093/aob/mcm052

[26] Shrivastava P, Kumar R. Soil salinity: A serious environmental issue and plant growth promoting bacteria as one of the tools for its alleviation. Saudi Journal of Biological Sciences. 2015;**22**(2):123-131. DOI: 10.1016/j.sjbs.2014.12.001

[27] Munns R, Tester M. Mechanisms of salinity tolerance. Annual Review of Plant Biology. 2008;**59**:651-681. DOI: 10.1146/annurev.arplant.59.032607.092911

[28] Grieve CM, Grattan SR, Maas EV. Plant salt tolerance. In: Wallender WW, Tanji KK, editors. ASCE Manual and Reports on Engineering Practice. Vol. 71. 2012. pp. 405-459

[29] Kotula L, Garcia Caparros P, Zörb C, Colmer TD, Flowers TJ. Improving crop salt tolerance using transgenic approaches: An update and physiological analysis. Plant, Cell & Environment. 2020;**43**(12):2932-2956. DOI: 10.1111/pce.13865

[30] Panabokke CR. Rice soils of Sri Lanka. In: Soils and Rice. Los Banos, Philippines: IRRI; 1978. pp. 15-23

[31] Department of Census and Statistics (DCS), Sri Lanka Statistical Abstract, Agriculture and Environment Statistics Division Department Census and Statistics. 2021. Available on: http://www.statistics.gov.lk/

[32] Sirisena DN, Wanigasundera WADP. Soil Salinity in Paddy Fields of Sri Lanka and Best Practices to Avoid Crop Failures Due to Soil Salinity. Agricultural Extension in South Asia. 2021. Available from: https://www.aesanetwork.org/blog-165-soil-salinity-in-paddy-fields-of-sri-lanka-and-best-practices-to-avoid-crop-failures-due-to-soil-salinity/

[33] National Agriculture Information and Communication Centre (NAICC). 2022. Available from: https://doa.gov.lk/naicc-publications-crop-calender/

[34] Marambe B, Punyawardena R, Silva P, Premalal S, Rathnabharathie V, Kekulandala B, et al. Climate, climate risk, and food security in Sri Lanka: The need for strengthening adaptation strategies. In: Handbook of Climate Change Adaptation. Berlin, Heidelberg: Springer; 2015. p. 1759-1789. DOI: 10.1007/978-3-642-38670-1_120

[35] Department of Meteorology (DOM). 2016. Available from: http://www.meteo.gov.lk/index.php?option=com_content&view=article&id=94&Itemid=310&lang=en

[36] Domroes M. The Agroclimate of Ceylon. A Contribution Towards the Ecology of Tropical Crops. Wiesbaden: Steiner; 1974

[37] Merrey DJ, Rao PS, Martin E. Irrigation management research in Sri Lanka: A review of selected literature. In: IIM1 Occasional Paper. Colombo, Sri Lanka: International Irrigation Management Institute; 1988

[38] Sivayoganathan C, Mowjood MIM. Role of extension in irrigation water management in Sri Lanka. Tropical Agricultural Research and Extension. 2003;**6**:49-59. DOI:10.4038/tare.v6i0.5439

[39] Jayawardane SN. Socio-Economic Constraints and Future Prospects for Crop Diversification in Minor Irrigation Schemes. Colombo, Sri Lanka: Workshop on crop diversification; 1996

[40] Chandrasiri S, Galagedara L, Mowjood M. Impacts of rainfall variability on paddy production: A case from Bayawa minor irrigation tank in Sri Lanka. Paddy and Water Environment. 2020;**18**:443-454. DOI: 10.1007/s10333-020-00793-9

[41] Gopalakrishnan T, Kumar L. Modeling and mapping of soil salinity and its impact on Paddy lands in Jaffna peninsula, Sri Lanka. Sustainability. 2020;**12**(20):8317

[42] Gopalakrishnan T, Kumar L, Mikunthan T. Assessment of spatial and temporal trend of groundwater salinity in Jaffna peninsula and its link to paddy land abandonment. Sustainability. 2020; **12**(9):3681

[43] Gopalakrishnan T, Kumar L. Linking long-term changes in soil salinity to paddy land abandonment in Jaffna peninsula,

Sri Lanka. Agriculture. 2021;**11**(3):211. DOI: 10.3390/agriculture11030211

[44] Abearathne MS, Hettiarachchi K, Chithrapala NHMS, Samarakoon Manike SM. Annals of Sri Lanka Department of Agriculture. 2013;**15**: 100-108

[45] Perera MD, Ranasinghe TK, Piyadasa RU, Jayasinghe GY. Risk of seawater intrusion on coastal community of Bentota river basin Sri Lanka. Procedia Engineering. 2018;**212**:699-706. DOI: 10.1016/j.proeng.2018.01.090

[46] Ranasinghe TKGP, Piyadasa RUK. Optimising usage of salinized lands in the lower part of the river basin for the coastal community in bentota, Sri Lanka. Journal of the National Science Foundation of Sri Lanka. 2020;**48**:379-396. DOI: 10.4038/jnsfsr.v48i4.8814

[47] Subasinghe S. Development of salt-affected lands in southern Sri Lanka with special reference to Hambantota district. Geography. 2004;**1**:1-7

[48] Lokupitiya EY, Agrawal M, Ahamed T, Mustafa N, Ahmed B, Vathani A, et al. Evaluation of best management practices with greenhouse gas benefits for salt-affected paddy soils in South Asia. APN Science Bulletin. 2020;**10**(1):41-49. DOI: 10.30852/sb.2020.1042

[49] Dharmasena TH, Piyadasa RU, Weerasinghe KD, Liyanage JA, Weerasinghe WA. Soil and groundwater salinity variation in Matara district coastal belt-Sri Lanka. In: Proceedings of the 64th Annual Sessions of Sri Lanka Association for the Advancement of Science, Colombo, Sri Lanka. 2008

[50] Rathnayake WMUK, Sirisena DN, Wanninayake WMN. Assessment of temporal variation of soil salinity in paddy fields In Puttalam District. Environmental Earth Sciences. 2015;**1**: 124-132

[51] Piyadasa RUK. River sand mining and associated environmental problems in Sri Lanka. IAHS-AISH Publication. 2011;**349**:148-153

[52] Rathnayake W, De Silva R, Dayawansa N. Assessment of the suitability of temperature and relative humidity for rice cultivation in rain fed lowland paddy fields in Kurunegala district. Tropical Agriculture. 2016;**27**: 370-388

[53] Department of Agriculture (DOA). Paddy Soil Fertility Maps. 2022. Available from: https://doa.gov.lk/soil-fertility-map/

[54] Dulanjalee PHE, Pitawala A. Salt mineralogy of soils in the area of Nawagatnthegama, Sri Lanka. Sri Lanka: Proceedings of the Peradeniya University Research Sessions; 2008;**13** (1):397-399

[55] Sirisena DN. Development of rice production techniques for increase of self-sufficiency of staple food in Sri Lanka. In: Proceedings of the Evaluation Meeting for AFACI Pan Asia Projects on Development of Rice Production Techniques for Increase of Self-Sufficiency of Staple Food in Asia. Bangkok, Thailand: Thir; 2013

[56] Clermont-Dauphin C, Suwannang N, Grünberge O, Hammecker C, Maeght JL. Yield of rice under water and soil salinity risks in farmers' fields in Northeast Thailand. Field Crops Research. 2010;**118**(3): 289-296. DOI: 10.1016/j.fcr.2010.06.009

[57] Krishnasamy K, Bell R, Ma Q. Wheat responses to sodium vary with potassium

use efficiency of cultivars. Frontiers in Plant Science. 2014;**5**(11):1-10. DOI: 10.3389/fpls.2014.00631

[58] Sirisena DN, Jayawardena NP, Wickramasinghe WMADB, Rathnayake WMUK, Mapa RB. Spatial variability of some soil fertility parameters in rice fields of Girithale irrigation scheme. Annals of Sri Lanka Department of Agriculture. 2008;**10**: 199-210

[59] Sirisena DN, Rathnayake WMUK, Herath HME. Productivity enhancement of saline paddy soils in Angiththamkulama yaya, Sri Lanka: A case study. Pedologist. 2010;**53**(3): 96-100

[60] Bandara WMJ, Kumaragamage D, Wickramasinghe DB, Weerawarne SBA. A site specific fertilizer management strategy to increase the rice yield in low country intermediate zone. Journal of Soil Science Society of Sri Lanka. 2005; **17**:32-44 https://dl.nsf.gov.lk/handle/1/16303

[61] Panabokke CR, Nagarajah S. The soil fertility characteristics of the Rice growing soils of Ceylon. Tropical Agriculturist. 1964;**CXX**:1-30

[62] Dodermann A, Fairhurst T. Rice Nutrient Disorders and Nutrient Management. Handbook Series. Singapore: Potash & Phosphate Institute of Canada (PPIC), East and Southeast Asian Programs; and Makati City, Philippines: International Rice Research Institute (IRRI); 2000. ISBN 981-04-2742-5

[63] Opatha KN, Lokupitiya E. Study of salinity levels and impact of saltwater intrusion on coastal paddy areas of wet zone of Sri Lanka. In: Proceedings of Symposium on Agrobiodiversity for Climate Change Adaptation, Food and Nutrition (Gannoruwa: Agriculture Publication Unit, Department of Agriculture). Food and Nutrition. 2019. pp. 71

[64] Lin L, Pussella P. Assessment of vulnerability for coastal erosion with GIS and AHP techniques case study: Southern coastline of Sri Lanka. Natural Resource Modeling. 2017;**30**(4):e12146. DOI: 10.1111/nrm.12146

[65] Ministry of Land. National Report on Desertification/Land Degradation Sri Lanka (Pdf). Sri Lanka: Ministry of Land; 2000 Available at: http://www.unccd-prais.com/Uploads/GetReportPdf/e9951953-8b60-41b4-a8fca0fa014a4a8e

[66] Edirisinghe EANV, Karunarathne GRR, Samarakoon ASMNB, Pitawala HMTGA, Dharmagunawardhane A, Tilakarathna IANDP. Assessing causes of quality deterioration of groundwater in Puttalam, Sri Lanka, using isotope and hydrochemical tools. Isotopes in Environmental and Health Studies. 2016; **52**:4-5. DOI: 10.1080/10256016.2015.1127918

[67] Mathiventhan T, Gorman D, Jayasingam T. Sea-level rise, coastal salinity and vegetation changes in Sri Lanka. Philosophical Transactions of the Royal Society. 2022;**380**(2221): 20210142. DOI: 10.1098/rsta.2021.0142

[68] Senanayake RMNH, Herath HMVG, Wickramesinghe IP, Udawela UAKS, Sirisena DN. Phenotypic screening of Rice varieties for tolerant to salt stress at seed germination, seedling and maturity stages. Tropical Agricultural Research. 2017;**29**(1):90-100. DOI: 10.4038/tar.v29i1.8300

[69] Sugirtharan M, Pathmarajah S, Mowjood MIM. Spatial and temporal

dynamics of water quality in Batticaloa lagoon in Sri Lanka. Tropical Agricultural Research. 2017;**28**(3):281-397. DOI: 10.4038/tar.v28i3.8232

[70] Kendaragama KMA. Salinity status of agro-well water in irrigated areas of the dry zone. Krushi. 1997;**16**:2-3

[71] MHJP G, Nadeeka K. Rainfall trends in Anuradhapura: Rainfall analysis for agricultural planning. RUSL Journals. 2013;**1**:38-44

[72] Silva EIL, Katupotha J, Amarasinghe O, Manthrithilake H, Ariyaratna R. Lagoons of Sri Lanka: From the Origins to the Present. Colombo, Sri Lanka: International Water Management Institute (IWMI); 2013. p. 122. DOI: 10.5337/2013.215

[73] Dharmasena TH, Piyadasa RU, Weerasinghe KD, Liyanage JA, Weerasinghe WA. Soil and groundwater salinity variation in Matara district coastal belt-Sri Lanka. In: Proceedings of the 64th Annual Sessions of Sri Lanka Association for the Advancement of Science, Colombo, Sri Lanka. 2008. p. 282

[74] Chen T, Zhu Y, Chen K, Shen C, Zhao X, Shabala S, et al. Identification of new QTL for salt tolerance from rice variety Pokkali. Journal of Agronomy and Crop Science. 2020;**206**(2):202-213. DOI: 10.1111/jac.12387

[75] Prusty MR, Kim SR, Vinarao R, Entila F, Egdane J, Diaz MGQ, et al. Newly identified wild rice accessions conferring high salt tolerance might use a tissue tolerance mechanism in leaf. Frontiers in Plant Science. 2018;**9**:417. DOI: 10.3389/fpls.2018.00417

[76] Costa WAJM, Wijeratne MAD, de Costa DM, Zahra ARF. Determination of the appropriate level of salinity for screening of hydroponically grown rice for salt tolerance. Journal of National Science Foundation Sri Lanka. 2012; **40**(2):123-136. DOI: 10.4038/jnsfsr.v40i2.4440

[77] Akbar M, Khush GS, Hilleris-Lambers D. Genetics of salt tolerance in rice. In: Proceedings of the International Rice Genetics Symposium. 1986. pp. 399-409

[78] Sirisena DN, Hemachandra PV. Screening rice varieties for salinity tolerance; comparison of different methods. Annals of Sri Lanka Department of Agriculture. 2007;**9**: 169-177

[79] Sala M, Shanthi P, Selvi B, Ravi V. Marker based genetic diversity of rice genotypes for salinity tolerance at panicle initiation stage. Vegetos. 2016; **29**(1):10-15. Available from: http://www.indianjournals.com/ijor.as

[80] International Rice Research Institute Breeding for Salt Tolerance in Rice. International Rice Research Institute. 2011. Available from: http://www.knowledgebank.irri.org/ricebreedingcourse/Breeding_for_salt_tolerance.htm

[81] Lakra WS, Gopalakrishnan A. Blue revolution in India: Status and future perspectives. Indian Journal of Fisheries. 2021;**68**(1):137-150. DOI: 10.21077/ijf.2021.68.1.109283-19

[82] Riyazuddin R, Verma R, Singh K, Nisha N, Keisham M, Bhati KK, et al. Ethylene: A master regulator of salinity stress tolerance in plants. Biomolecules. 2020;**10**(6):959. DOI: 10.3390/biom10060959

[83] Makihara D, Tsuda M, Morita M, Hirai Y, Kuroda T. Effect of salinity on the growth and development of rice (Oryza sativa L.) varieties. Japanese.

Journal of Tropical Agriculture. 1999; **43**(4):285-294. DOI: 10.11248/jsta1957.43.285

[84] Ranathunge K, Steudle E, Lafitte R. Control of water uptake by rice (Oryza sativa L.): Role of the outer part of the root. Planta. 2003;**217**(2):193-205. DOI: 10.1007/s00425-003-0984-9

[85] Krishnamurthy P, Ranathunge K, Nayak S, Schreiber L, Mathew MK. Root apoplastic barriers block Na+ transport to shoots in rice (Oryza sativa L.). Journal of Experimental Botany. 2011; **62**(12):4215-4228. DOI: 10.1093/jxb/err135

[86] Lakra N, Kaur C, Anwar K, Singla-Pareek SL, Pareek A. Proteomics of contrasting rice genotypes: Identification of potential targets for raising crops for saline environment. Plant, Cell & Environment. 2018;**41**(5):947-969. DOI: 10.1111/pce.12946

[87] Mishra M, Wungrampha S, Kumar G, Singla-Pareek SL, Pareek A. How do rice seedlings of landrace Pokkali survive in saline fields after transplantation? Physiology, biochemistry, and photosynthesis. Photosynthesis Research. 2021;**150**:117-135. DOI: 10.1007/s11120-020-00771-6

[88] Gregorio GB, Senadhira D. Genetic analysis of salinity tolerance in rice (Oryza sativa L.). Theoretical Applied Genetics. 1993;**86**:333-338. DOI: 10.1007/BF00222098

[89] Causse MA, Fulton TM, Cho YG, Ahn SN, Chunwongse J, Wu K, et al. Saturated molecular map of the rice genome based on an interspecific backcross population. Genetics. 1994; **138**(4):1251-1274. DOI: 10.1093/genetics/138.4.1251

[90] Zhang R, Rahman S, Vance GF, Munn LC. Geostatistical analyses of trace elements in soils and plants. Soil Science. 1995;**1**:37-43

[91] Ding HY, Zhang Y, Guo Y, Chen SL, Chen SY. RAPD tagging of a salt tolerant gene in rice. Chinese Science Bulletin. 1999;**43**:330-332. DOI: 10.1007/BF02884550

[92] Quan R, Wang J, Hui J, Bai H, Lyu X, Zhu Y, et al. Improvement of salt tolerance using wild rice genes. Frontiers in Plant Science. 2018;**8**:2269. DOI: 10.3389/fpls.2017.02269

[93] Gregorio GB, Senadhira D, Mendoza RD. Screening rice for salinity tolerance. IRRI Discussion Paper Series. 1997;**22**(1):30 International Rice Research Institute, Manila, Philippines

[94] Bonilla PS, Dvorak J, Mackill D, Deal K, Gregorio G. RFLP and SSLP mapping of salinity tolerance genes in chromosome 1 of Rice (Oryza sativa L.) using recombinant inbred lines. The Philippine Agricultural Scientist. 2002; **85**(1):68-76

[95] Niones JM. Fine mapping of the salinity tolerance gene on chromosome 1 of rice (*Oryza sativa* L.), using near isogenic lines. MS dissertation. Laguna, Philippines: University of the Philippines Los Baños; 2004

[96] Thomson MJ, de Ocampo M, Egdane J, Rahman MA, Sajise AG, Adorada DL, et al. Characterizing the Saltol quantitative trait locus for salinity tolerance in rice. Rice. 2010;**3**(2):148-160. DOI: 10.1007/s12284-010-9053-8

[97] Waziri A, Kumar P, Purty RS. Saltol QTL and their role in salinity tolerance in

rice. Austin Journal of Biotechnology and Bioengineering. 2016;**3**(3):1067

[98] Ren ZH, Gao JP, Li LG, Cai XL, Huang W, Chao DY, et al. A rice quantitative trait locus for salt tolerance encodes a sodium transporter. Nature Genetics. 2005;**37**(10):1141-1146. DOI: 10.1038/ng1643

[99] Hauser F, Horie T. A conserved primary salt tolerance mechanism mediated by HKT transporters: A mechanism for sodium exclusion and maintenance of high K+/Na+ ratio in leaves during salinity stress. Plant, Cell & Environment. 2010;**33**(4):552-565. DOI: 10.1111/j.1365-3040.2009.02056.x

[100] Platten JD, Egdane JA, Ismail AM. Salinity tolerance, Na+ exclusion and allele mining of HKT1; 5 in Oryza sativa and O. Glaberrima: Many sources, many genes, one mechanism? BMC Plant Biology. 2013;**13**:1-16. DOI: 10.1186/1471-2229-13-32

[101] Pulipati S, Somasundaram S, Rana N, Kumaresan K, Shafi M, Civáň P, et al. Diversity of sodium transporter HKT1; 5 in genus oryza. Rice Science. 2022;**29**(1):31-46. DOI: 10.1016/j.rsci.2021.12.003

[102] Lin HX, Zhu MZ, Yano M, Gao JP, Liang ZW, Su WA, et al. QTLs for Na+ and K+ uptake of the shoots and roots controlling rice salt tolerance. Theoretical and Applied Genetics. 2004;**108**:253-260. DOI: 10.1007/s00122-003-1421-y

[103] Abeysiriwardena DS, De Z. A simple screening technique for salinity tolerance in rice: Germination rate under stress. International Rice Research Notes. 2004;**29**(2):78-79

[104] Subasinghe A, Nissanka NAASP, Weerakoon WMW. Identification of salt tolerant rice varieties at the seed germination stage and its relationship to seed husk thickness and Ion absorption. Tropical Agricultural Research. 2007;**19**(1):219-228

[105] Safeena MIS, Bandara, DC, & Sumanasinghe VA. Effect of Salt Stress on Leaf Area Rate of Photosynthesis and Proline Accumulation of Rice Varieties at Different Growth Stages. 2002. https://dl.nsf.gov.lk/handle/1/11885

[106] Dahanayaka BA, Gimhani DR, Kottearachchi NS, Samarasighe WLG. QTL mapping for salinity tolerance using an elite rice (Oryza sativa) breeding population. SABRAO Journal of Breeding & Genetics. 2017;**49**(2):123-124

[107] Rajapakse RMT, Sandanayake CA. & Pathinayaka BD. Foot Print in Rice Variety Improvement and its Impact on Rice Production in Sri Lanka. 2010. Available from: http://www.goviya.lk/agri_learning/Paddy/Paddy_Reserach/Paddy_pdf/V10.pdf

[108] Department of Census and Statistics (DCS) Statistical Abstract. Agriculture and Environment Statistics Division Department Census and Statistics, Sri Lanka. 2022. Available from: http://www.statistics.gov.lk/Agriculture/StaticalInformation/rubpaddy

[109] Najim MMM, Jayakody KPK. Salinity development in the dry zone of Sri Lanka – a review. In: Mohammad K, Mohammad AK, editors. Crop and Forage Production Using Saline Waters. Delhi: Daya Publishing House; 2008. p. 319-328

[110] Bhumbla DR, Abrol IP. Is your water suitable for irrigation? Get it tested. Indian Farming. 1972;**22**:15-17

[111] Massoumi AM. Saline water classification in Iran by rational methods. In: Proceedings of the International Salinity Conference, 16-20 August 1976. Lubbock, Texas: Texas Tech University; 1976. p. 455-465

[112] Pla I, Dappo F. Field testing of a new system for qualifying irrigation water. In: Proceedings of International Conference on Managing Saline Water for Irrigation. Lubbock, Texas: Texas Tech. University; 1976. pp. 376-387

[113] Abrol IP, Palaniappan SP. Green manure crops in irrigated and rainfed lowland rice-based cropping systems in South Asia. In: Sustainable Agriculture: Green Manure in Rice Farming. Philippines: IRRI; 1988. pp. 71-82

[114] Ota K, Yasue T, Iwatsuk KM. Studies on the salt injury to crop: Relation between the salt injury and the pollen germination in rice. Research Bulletin of the Faculty of Agriculture, Gifu University. 1956;7:15-20

[115] Reddy PR, Goss JA. Effect of salinity on pollen I. Pollen viability as altered by increasing osmotic pressure with NaCl, MgCl2, and CaCl2. American Journal of Botany. 1971;**58**(8):721-725. DOI: 10.2307/2441469

[116] Abdullah ZUN, Ahmad R, Ahmed J. Salinity-induced changes in the reproductive physiology of wheat plants. Plant and cell Physiology. 1978;**19**(1): 99-106. DOI: 10.1093/oxfordjournals.pcp.a075584

[117] Khatun S, Flowers TJ. The effects of salinity, applied at different times, on the growth, grain yield and pollen viability of rice. Journal of Experimental Botany. 1994

[118] Ponnamperuma FN. Properties of Tropical Rice Soils. Mexico: Text of lecture at the Tropical Agriculture Collage; 1987

[119] Abdullah Z, Khan MA, Flowers TJ. Causes of sterility in rice under salinity stress. In: Ahmad R, Malik KA, editors. Prospects for Saline Agriculture. Tasks for Vegetation Science. Vol 37. Dordrecht: Springer; 2002. DOI: 10.1007/978-94-017-0067-2_19

[120] Rao PS, Mishra B, Gupta SR, Rathore A. Reproductive stage tolerance to salinity and alkalinity stresses in rice genotypes. Plant Breeding. 2008;**127**(3): 256-261. DOI: 10.1111/j.1439-0523.2007.01455.x

[121] Murty PSS, Murty KS. Spikelet sterility in relation to nitrogen and carbohydrate contents in rice. Indian Journal of Plant Physiology. 1982;**25**(1): 40-48

[122] Saini HS, Westgate ME. Reproductive development in grain crops during drought. Advances in Agronomy. 1999;**68**:59-96. DOI: 10.1016/S0065-2113(08)60843-3

[123] Rang ZW, Jagadish SVK, Zhou QM, Craufurd PQS, Heuer. Effect of high temperature and water stress on pollen germination and spikelet fertility in rice. Environmental and Experimental Botany. 2011;**70**(1):58-65. DOI: 10.1016/j.envexpbot.2010.08.009

[124] Yoshida S. Effects of temperature on growth of the rice plant (Oryza sativa L.) in a controlled environment. Soil Science and Plant Nutrition. 1973;**19**(4): 299-310. DOI: 10.1080/00380768.1973.10432599

[125] Jagadish SK, Craufurd PQ, Wheeler T. High temperature stress and spikelet fertility in rice (Oryza sativa L.). Journal of experimental botany. 2007;

58(7):1627-1635. DOI: 10.1093/jxb/erm003

[126] Weerakoon WMW, Abeywickrama T, De Costa WAJM, Maruyama A. Out-crossing of heat stress affected spikelets of lowland rice in the sub-humid zone of Sri Lanka and its long-term implications. In: Proceedings of the 3rd Annual MARCO Symposium, Tsukuba, Japan. 2009. pp. 5-7

[127] de Abeysiriwardena DSZ, Ohba K, Maruyama A. Influence of temperature and relative humidity on grain sterility in rice. Journal of National Science Foundation of Sri Lanka. 2002;**30**:33-41. DOI: 10.4038/jnsfsr.v30i1-2.2559

[128] Reddy INBL, Kim BK, Yoon IS, Kim KH, Kwon TR. Salt tolerance in rice: Focus on mechanisms and approaches. Rice Science. 2017;**24**(1):123-144

[129] Mainuddin M, Maniruzzaman MD, Alam MM, Mojid MA, Schmidt EJ, Islam MT, et al. Water usage and productivity of Boro rice at the field level and their impacts on the sustainable groundwater irrigation in the north-West Bangladesh. Agricultural Water Management. 2020; **240**:106-294. DOI: 10.1016/j.agwat.2020.106294

[130] Assefa Y, Yadav S, Mondal MK, Bhattacharya J, Parvin R, Sarker SR, et al. Crop diversification in rice-based systems in the polders of Bangladesh: Yield stability, profitability, and associated risk. Agricultural Systems. 2021;**187**:102986. DOI: 10.1016/j.agsy.2020.102986

[131] Yang Y, Zhou B, Feng L. Effects of different soil modifiers on salt improvement and distribution, crop growth of the gully land consolidation on loess plateau. Journal of Sensors. 2022;**1**(2022):1-17. DOI: 10.1155/2022/5282344

[132] Khamidov M, Ishchanov J, Hamidov A, Donmez C, Djumaboev K. Assessment of soil salinity changes under the climate change in the Khorezm region, Uzbekistan. International Journal of Environmental Research and Public Health. 2022;**19**(14):8794. DOI: 10.3390/ijerph19148794

[133] Fernando TK, Chandrapala L. Climate variability in Sri Lanka – A study of trends of air temperature, rainfall and thunder activity. In: Proceedings of the International Symposium on Climate and Life in Asia-Pacific. Brunei: Darussalam and University of Brunei, 1995. pp. 10-13

[134] Nissanka SP, Punyawardena BVR, Premalal KHMS, Thattil RO. Recent trends in annual and growing seasons' rainfall of Sri Lanka. In: Proceedings of the International Conference on the Impact of Climate Change in Agriculture. Mapalana: University of Ruhuna; 2011. pp. 249-263

[135] Premalal KHMS, Punyawardena BVR. Occurrence of extreme climatic events in Sri Lanka. In: Proceedings of the International Conference on Climate Change Impacts and Adaptations for Food and Environment Security, Peradeniya. 2013. pp. 49-57

[136] Ratnayake U, Herath S. Changing rainfall and its impact on landslides in Sri Lanka. Journal of mountain science. 2005;**2**:218-224

[137] Imbulana LALANI, Mollinga PP, Dixit A, Athukorala K. Water allocation between agriculture and hydropower: A case study of Kalthota irrigation scheme, Sri Lanka. In: Integrated Water Resources Management: Global Theory, Emerging Practice and Local Needs. 2006. pp. 219-248

[138] Wassmann R, Hien NX, Hoanh CT, Tuong TP. Sea level rise affecting the

Vietnamese Mekong Delta: Water elevation in the flood season and implications for rice production. Climatic Change. 2004;**66**:89-107. DOI: 10.1023/B:CLIM.0000043144.69736.b7

[139] Natcom. Democratic Socialist Republic of Sri Lanka, Initial National Communication under the United Nations Framework Convention on Climate Change. Padukka: State printing corporation; 2000

[140] Jayasiri HB, Dahanayaka DDGL. Saltwater intrusion and its impact on Koggala lagoon and associated waters, southern coast of Sri Lanka. Asian Journal of Water, Environment and Pollution. 2012;**9**(2):73-84

Chapter 3

Plant Adaptation to Salinity Stress: Significance of Major Metabolites

Maneesh Kumar, Himanshu Bharadwaj and Komal Kumari

Abstract

These genes increase the plant's tolerance to salt stress by producing proteins and metabolites that protect the cell against stress. More secondary metabolites including anthocyanins phenols, saponins, flavonoids, carotenoids, and lignins, etc., are produced by plants in salty conditions, but previous studies have only looked at a small portion of these compounds. Antioxidant activity and phenolic compound accumulation under salt stress have been linked in several studies. Proline accumulates in the cytoplasm and the vacuole, where it functions as an osmolyte and protects macromolecules against denaturation. Polyamines play a role in salt tolerance by regulating gene expression and ion flux. This means that metabolites are crucial for plant response to salt stress and maintaining agricultural productivity in salt-affected environments.

Keywords: abiotic factor, salinity, secondary metabolites, stress, plant defense

1. Introduction

The salinization of soil has emerged as one of the most significant environmental and socioeconomic problems on a global scale, and it is anticipated that this problem will become much more severe as a result of forecasted changes in the climate. It affects food security, water availability, and human health. Salinization of soil occurs when salt accumulates in the soil, making it unsuitable for crop growth. This can be caused by the accumulation of naturally occurring salts or by the introduction of salt-laden irrigation water. Salinization can also be caused by inadequate drainage, where water accumulates in the soil and dissolves and accumulates salts from the soil. Salinization can reduce crop yields and cause environmental damage such as increasing the salinity of nearby rivers and streams. To prevent salinization, farmers can use soil amendments to reduce the salt content of the soil and use irrigation practices that avoid the accumulation of salt in the soil. This can be caused by over-irrigation, poor drainage, and high levels of evaporation. The economic effects of salinization of soil can be seen in the form of reduced crop yields, increased. It is a process where the salt content in soil increases, which can endanger plant life and lead to soil erosion. Salinization is a result of increased precipitation, flooding, or irrigation that saturates the soil with salt. Salinization can also be the result of the chemical precipitation of salt from the atmosphere [1]. High levels of salt can cause soil structure degradation. It can lead to the formation of a hardpan layer, which can make it difficult for roots to penetrate the soil. This can further reduce the crop yields significantly. Soil salinity

IntechOpen

can have a significant adverse effect on crop yield. High levels of salinity can reduce crop yields by reducing germination rates, reducing plant growth, and increasing the susceptibility of plants to disease. Salinity can also reduce the availability of essential nutrients, resulting in stunted growth and reduced yields. Additionally, high salinity levels can result in increased water stress, which can further reduce crop yields. Recent estimates show that salt affects about 1125 million hectares of land around the world and that every year, 1.5 million hectares of land cannot be used for farming because the soil is too salty [2, 3]. It is hard to figure out how much agricultural production is lost, but it is thought that between 25% and 50% of all irrigated land is affected by salt accumulation, leading to decreased yields and the abandonment of agricultural land. It is estimated that up to 1 billion people in the world are affected by salinization and that the problem is growing. The World Bank has estimated that up to $20 billion a year is lost due to salinization [4]. Due to the extremely complex natural processes, abiotic stress inhibits plant development, reduces agricultural production, and further contributes to excessive soil degradation. Farmers' incomes and other local economies are solely slowed [1].

Plants make substances called secondary metabolites that help them do well in their environment. A variety of responses occur in plants and other species as a result of these tiny chemicals. They signal the continuation of perennial growth or the onset of dormancy, and they are responsible for triggering blooming, fruit set, and abscission. Depending on the circumstances, they can attract or repel microbes. There are over 50,000 distinct secondary metabolites present in plants. Secondary plant metabolites are responsible for the activities of therapeutic plants and a variety of modern pharmaceuticals. Secondary metabolites are produced by plants in response to environmental and biotic stimuli. These include defensive chemicals, attractants, and toxins. They are also responsible for plant growth, development, and health. Universities and drug companies are always looking for new secondary products in plants in the hopes of finding new products or, even better, new ways to treat diseases. As secondary metabolites, once-held hopes that better appreciating natural product distribution might aid in plant taxonomy were also entertained. This supplementary rationale is no longer relevant because plant classification is increasingly done by comparing DNA sequences [5, 6]. In this chapter, we talk in-depth about the fundamental contribution of secondary metabolites that enable plants to sustain soil salinity, increase agricultural production, and have no negative effects on the economy or people's health.

1.1 Soil salinity causing factors

Soil salinization can be caused by a number of factors, but most commonly it is due to improper irrigation practices that allow salts to accumulate in the soil. Other causes of salinization include soil type, water table fluctuations, and high evaporation rates. In some cases, poor agricultural practices, such as excessive fertilization, can also lead to soil salinization. In order to prevent soil salinization, it is important to practice proper irrigation techniques, ensure that the soil has adequate drainage, and make sure to use the appropriate amount of fertilizer for the crops being grown. Additionally, certain soil amendments, such as gypsum, can help reduce the number of salts in the soil [2, 7]. One of the most common and bad effects of soil flooding is that it makes the soil saltier. Salinity is based on how much salt is dissolved in the soil, and it can be changed by a number of things. Poor drainage can also make it easier for certain types of weeds to grow, which can lead to soil erosion and other problems. It can also cause the soil to get harder, which can make it harder for plants to get enough oxygen and nutrients. Soil salinization occurs when seawater or other salt water floods a field or other piece of land. When water evaporates, salt and other minerals are left behind. These can build up in the soil and make it less fertile. This can cause long-term problems with soil salinity in places that flood often [7]. To mitigate the effects of soil salinity caused by flooding, farmers and landowners should improve drainage, plant salt-tolerant crops, and restrict the amount of irrigation water permitted to pool on the land. When soil is inundated, water from rivers, streams, and other sources can carry enormous quantities of dissolved salts. High concentrations of nutrients, such as nitrogen and phosphorus, can also contribute to an increase in salinity. Inadequate drainage can also raise the water table, leading to flooding and other water-related problems [8, 9]. Salt can build up in the soil as a result of over-irrigation. This can cause a variety of issues, including decreased crop yields, soil structure damage, and plant toxicity. Salt buildup can also cause water logging, which reduces soil oxygen levels and damages plant roots. Farmers should employ water-saving irrigation techniques, such as drip irrigation, to mitigate the impacts of soil salinity caused by over-irrigation. Farmers should also utilize soil tests to detect the salinity of the soil and alter their irrigation operations accordingly. Finally, the use of soil supplements, such as gypsum, can aid in the reduction of salt accumulation in the soil [10]. Climate change and natural disasters (such as a tsunami) that make the soil saltier are still problems for agriculture. When the Earth's temperature rises, the water that falls as rain or snow becomes increasingly saline. An increase in saline water can dissolve salts in the soil, making plant growth more challenging. Also, when the Earth's seas warm, more salt is released into the atmosphere. This rise in salinity can kill marine life, making it impossible for them to live. Finally, as the Earth's surface dries up, salts collect in the soil more easily [11]. Drought, excessive salinity, and cold temperatures are all climatic factors that have a negative impact on plant growth and crop yield. The growth of a plant and the production of secondary metabolites are both affected by environmental conditions such as temperature, humidity, and the intensity of the light, as well as the availability of water, minerals, and carbon dioxide. Cellular dehydration brought on by exposure to salt generates osmotic stress, which in turn leads to the loss of cytoplasmic water and a consequent shrinking of the cytosol and vacuoles. Depending on the severity of the salt stress, plants may either accumulate or decrease a number of different secondary metabolites [12]. Generally, salt-stressed plants are known to accumulate a wide range of secondary metabolites, including polyamines, polyphenols, and flavonoids, which act as antioxidants and chelators of metal ions, thereby helping to protect the plants from oxidative damage caused by the salt stress. In addition, the accumulation of these compounds can also help to increase osmotic pressure, allowing plants to cope with the osmotic stress caused by salt. On the other hand, plants under salt stress may also decrease the synthesis of certain secondary metabolites, such as terpenes and alkaloids, which can be toxic to plants when present in high concentrations. Therefore, the accumulation or decrease of secondary metabolites in salt-stressed plants is largely dependent on the severity of the salt stress.

1.2 Plant defenses against salt stress and metabolic alertness

There are several different ways in which plants react when they are subjected to salt stress. The physiological and the biochemical are the two categories that can be used to classify these responses (**Figure 1**). Alterations in the growth rate of the plant are one type of physiological reaction. Other types of physiological reactions include changes in photosynthesis, respiration, transpiration, water uptake, nutrient uptake,

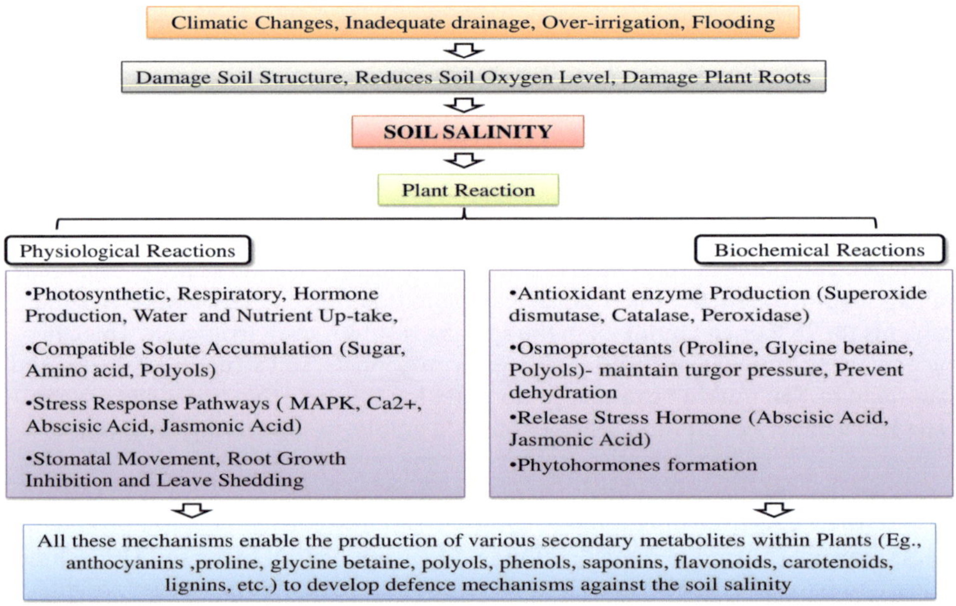

Figure 1.
Salt stress and metabolic reactions are physiological and biochemical: salt stress intensity affects secondary metabolite accumulation in salt-stressed plants. Salt stress increases osmotic potential, decreases water potential, increases photosynthetic rate, and decreases stomatal conductance. Salt stress increases antioxidant enzyme activity, stress-related gene expression, proline buildup, and soluble sugar accumulation. Plants need these responses to stay healthy and avoid salt stress.

and hormone production. Plants respond to salt stress by increasing the concentration of compatible solutes in their cells, which helps them maintain their turgor pressure and prevent dehydration [13]. Compatible solutes are small molecules that can be accumulated in the cell without disrupting its osmotic balance. Examples of compatible solutes include sugars, amino acids, and polyols. Plants also respond to salt stress by activating various stress-response pathways, such as the MAPK, calcium, and ABA pathways, which help them cope with the stress and protect them from damage [14]. Stomatal movement also gets affected under this salt stress. Plants close their stomata to reduce the amount of water lost through transpiration. This helps them conserve water and reduce the amount of salt that enters the plant. Stomata are small pores on the surface of leaves that open and close to regulate the exchange of gases and water vapor. When the stomata are closed, the plant reduces the amount of water vapor that is released into the atmosphere. This helps the plant conserve water and reduce the amount of salt that enters the plant [15, 16]. In some cases, plants may shed their leaves and can cause root growth inhibition, which reduces the amount of water and nutrients taken up by the plant to cut the amount of salt uptake. Salt stress led to antioxidant enzymes (such as superoxide dismutase, catalase, and peroxidase), osmoprotectants (such as proline, glycine betaine, and polyols to help them maintain their turgor pressure and prevent dehydration), stress hormones (abscisic acid and jasmonic acid), and phytohormones (gibberellins and cytokinins) production [17, 18]. Other types of physiological responses include changes in the structure of the plant such as an increase in the production of root hairs and thicker cell walls. Changes in the plant's metabolism, such as the creation of enzymes and other proteins that assist the plant in coping with salt stress, are examples of biochemical responses that take place as a result of salt stress. In addition, plants may create compounds that assist them in absorbing and storing salt, or they may develop compounds that assist them in excreting excess salt as a defense mechanism against the effects of salt stress.

Plants can not get away from potentially dangerous situations, so they have evolved to have different defense mechanisms to protect themselves. These include physical barriers such as thorns, chemical defenses such as toxins and poisons, and camouflage to blend in with their environment. Some plants also produce chemicals that attract predators of the herbivores that would otherwise feed on them. Some plants, such as certain species of grasses, are able to tolerate high levels of salinity. Other plants may be able to survive in the short term but suffer long-term damage. In extreme cases, the plants may die. There are several strategies that can be employed to help plants cope with high levels of salinity such as using salt-tolerant varieties of plants, avoiding over-irrigation, and using soil amendments to improve the soil's structure and drainage [19]. Metabolites are used for a variety of purposes in plants. Plants can use these metabolites to help regulate their internal water balance, allowing them to cope with salt stress [19, 20]. Such compounds help plants detoxify and eliminate excess salt from their cells. They can be used to protect the plant's cells from the damaging effects of salt. They can be used as a source of energy, to help regulate growth and development, to produce hormones, to protect against environmental stress, to produce pigments, to aid in defense against pathogens, and to help in the synthesis of other molecules [21]. Some plants have evolved mechanisms to cope with soil salinity. These mechanisms involve the production of metabolites, such as proline, glycine betaine, and trehalose, which help the plant to maintain its water balance and protect its cells from the damaging effects of salt. Additionally, some plants produce special root structures that help to reduce the uptake of salt from the soil.

The prospects of metabolic alertness in crop production against salinity stress are very promising. By understanding the role of metabolic alertness in the response to salt stress, researchers can develop new strategies to improve crop performance under salinity. It refers to the ability of a crop to detect and respond to salt stress. It could improve the drought tolerance of crops by reducing the amount of water lost through transpiration. It helps crops adjust their metabolic processes to adapt to the salinity environment. Various studies have shown that crops with higher metabolic alertness respond better to salinity stress than crops with lower alertness. This is done by increasing the production of certain proteins that are involved in salt tolerance. It also involves the upregulation of certain genes that are involved in salt tolerance. Finally, metabolic alertness can be used to improve the nutrient content of crops by increasing the uptake of minerals and other nutrients.

2. Secondary metabolic profile of plants under soil salinity

The secondary metabolic profile of plants in soils under saline conditions shows an increase in the activity of enzymes involved in the synthesis of secondary metabolites. These enzymes are typically more active when the concentration of salts is high. In addition, plants may upregulate the expression of genes encoding other enzymes involved in the synthesis of secondary metabolites in response to salt stress. Several stressors are imposed on plants, such as the presence of certain elicitors or signal molecules, resulting in the accumulation of secondary metabolites [21, 22]. It is possible for the secondary metabolites to be created in reaction to a wide variety of stimuli such as being injured, experiencing extreme cold or heat, dehydration, or exposure to salt or light. These compounds are also capable of being created as a result of the metabolic response that the plant has developed in reaction to being under stress. It is common for secondary metabolites to play a role in a plant's defensive mechanism, whether it is against herbivores or pathogens. They are also able to contribute to the communication between plants, both with one another and with their surrounding environment.

In general, soils that are dry and infertile are more likely to become salinized since they are less able to resist the accumulation of salts. Salinity can also increase as a result of climate change as warmer temperatures allow more water to evaporate, leaving behind more salts. The inability of plants to absorb water due to an overabundance of salts is one of the many reasons why salinization is a resource concern that threatens agricultural output. This soil salinity includes sodium (Na^+), potassium (K^+), chloride (Cl^-), and sulfate (SO_4^{2-}). Ions, such as Na^+ and Cl^-,are not utilized by plants for nourishment; however, ions, such as K^+ and SO_4^{2-}, are. As a consequence of this, sodium ions and chloride ions are regularly investigated about soil salinity. The term "sodicity" refers to the degree to which the concentration of Na^+ in the soil solution rises in comparison to the concentration of other exchangeable cations [23]. Indirectly, through the worsening of soil physical conditions, salinity and sodicity affect plant growth through their impacts on the water intake by plants, the availability of nutrients for plants, and by imposing plant toxicity. These effects can be attributed to salinity and sodicity's respective impacts on water intake by plants, nutrient availability for plants, and plant toxicity. Both natural processes and human-caused interventions can contribute to an increase in salinity [24, 25]. The situation might become severe when salt contributions to the soil surface become too high. It exclusively hampers seed germination and plant growth. Alternatively, sodium (Na^+) may cause the soil to become more dispersed. The electrical conductivity (EC) of the ground is used to assess the dangers posed by salt, whereas the exchangeable sodium percentage (ESP) is used to assess the dangers posed by sodium. By making EC more concentrated, salt dispersion may affect the possible damage [24]. Salts will nearly usually first occur at lower elevations on a landscape, and they will gradually ascend to higher heights through time. This phenomenon is occurring on millions of hectares worldwide, including in Australia, Canada, Montana, Minnesota, and North Dakota and South Dakota. It also occurs in other nations, including New Zealand and Australia. There is a correlation between a decline in agricultural yields and profitability and an increase in soil electrical conductivity (EC) in locations where salt has impacted the soil [26].

2.1 Plant metabolites and its significance in environment

Plants store many compounds or "specialized metabolites." These tiny compounds affect plants and other living things. They blossom, fruit, then abscise or retain everlasting growth. Beside fighting bacteria, these compounds can also draw in or ward off pests. These substances are referred to as "secondary metabolites" [27]. These chemicals help a creature adapt to environmental changes and interact with other organisms. They protect against viruses, pests, and herbivores, respond to environmental stress and connect organisms. So far, about 50,000 secondary metabolites that come from plants have been studied. Plant secondary metabolites are the main therapeutic agents in ancient and modern medicines. It is the goal of many academic and pharmaceutical institutions to discover new goods or, better yet, new treatment approaches for a wide range of disorders by doing extensive research into the plant's secondary compounds. Once upon a time, it was thought that understanding the spread of natural products would help classify plants [26, 28]. During a plant's life, as it interacts with its complicated multi-kingdom microbiome, which is made up of both good and bad microbes, these specialized metabolites have been shown to play one of the most important and noticeable roles. Plant microbiomes are also in charge of controlling how a plant's metabolism works. Because of this, plant microbiomes are involved in a lot of the things listed above. A lot of plant secondary metabolites are important to the economy because they are good for human health [29, 30] and help increase the amount of food that can be grown. Even though several protein-substrate and protein-metabolite complexes have been identified, the majority of their biological functions remain unconfirmed [31]. Scientists have identified functioning secondary metabolites and metabolic pathways in plants using metabolomics. These findings apply to both fundamental and applied research. Nuclear magnetic resonance spectroscopy, Fourier transform near-infrared spectroscopy, capillary electrophoresis mass spectrometry, gas chromatography-MS, liquid chromatography-MS, MS imaging, and live single-cell MS are some of the most prevalent methods (LSC-MS). These approaches generally work together since they examine distinct metabolites. These techniques aid researchers in gaining a more comprehensive understanding of how metabolic networks in plants are regulated under different biotic and abiotic circumstances.

The plant microbiome helps the plant fight disease [32], get nutrients [33], and protect itself from living and nonliving environmental threats [34, 35]. Large-scale parallel sequencing enhanced plant microbiome research 15 years ago. These studies discovered plant microbiomes and interactions. Seed, core, synthetic community, defensive, and epiphytic microbiomes are some of the examples. Plant microbiomes respond differently to biotic and abiotic stimuli. More evidence reveals that complex feedback loops between plants, microorganisms, and their physical and chemical environments shape plant microbiomes. Genomic and molecular biology developments make studying plant specialized metabolites' biosynthesis pathways' structural and regulatory components easier. These methods can also be used to develop and test synthesizing-deficient mutant strains [6, 36]. Apart from these environmental elements such as light, temperature, soil water, soil quality, and salinity all have a significant role in the secondary metabolites' ability to accumulate. Changes in a single environmental element, even when all others are held constant, can affect the levels of secondary metabolites in most plant species. Plants are sensitive to the ionic or osmotic pressure induced by salinity, which can either promote or decrease the accumulation of particular secondary metabolites. Their secondary metabolites provide protection of the plant cells from the oxidative damage produced by ion accumulation at the cellular and subcellular levels; salt stress may operate as an elicitor of secondary metabolites, which mitigate the harmful effects of salinity [37].

2.2 Secondary metabolic profiles decrease with salinity stress

As salinity increases, secondary metabolic profiles decrease in a variety of species. This can be seen in the decreased levels of metabolites such as carbohydrates, lipids, and proteins. This decrease in metabolic activity can be due to a number of factors, including a decrease in the number of cells and a decrease in the activity of metabolic enzymes. An increase in soil saltiness and ion accumulation represents one of the important abiotic factors that adversely affect the growth and production of cultivated plants. High NaCl concentrations impede plant growth due to a decrease in hydraulic conductivity (hyperosmotic stress) and the accumulation of ions to harmful levels for their proliferation (hypertonic stress). Plants change their biochemical and physiological processes in response to these stresses. Observing gene regulation, production of functional proteins, and accumulation of tiny molecules (i.e., metabolites) has allowed researchers to concentrate on plant signal perception and adaptability to an unfavorable environment [38, 39]. Salinity stress can have various effects on plants, including changes in their secondary metabolism. Secondary metabolites are compounds produced by plants that are not directly involved in growth and development but instead play roles in various functions such as defense against herbivores, attraction of pollinators, and communication with other organisms [26, 40].

Several studies have reported that salinity stress can lead to a decrease in the production of secondary metabolites in plants. For example, a study was conducted on exposure of basil plants to salt stress resulted in a decrease in the levels of certain secondary metabolites, including β-carotene, cryptoxanthin, lutein flavonoids, and phenolic acids in the leaves and flowers. Salinity stress has been shown to decrease the levels of certain secondary metabolites in basil plants [41]. Another study reported that salinity stress reduced the levels of secondary metabolites in grapevine leaves, including stilbenoids and flavonoids [42]. However, it is important to note that the effect of salinity stress on secondary metabolism can vary depending on the plant species and the specific metabolites involved. Some studies have also reported an increase in the production of certain secondary metabolites in response to salinity stress, suggesting that the relationship between salinity stress and secondary metabolism is complex and not fully understood. As a result of salinity stress, plants produce secondary metabolites that help them to adapt to the new environment. These metabolites can help plants reduce water loss, increase their resistance to pests and pathogens, and increase their salt tolerance.

Several studies have shown that sugars, amino acids, and organic compounds, which are primary metabolites, play a role in osmotic regulation. In contrast, secondary metabolites, which are the final products of primary metabolites, are more species-specific and are connected with plant protection due to their numerous roles (e.g., serving as antioxidant activity, superoxide radical (ROS) scavengers, and regulatory molecules) [43]. Even though the production of secondary metabolites such as phenols, saponins, flavonoids, carotenoids, and lignins, etc. usually goes up in salt-stressed plant, only a few target compounds have been looked at in detail in previous studies [44, 45]. Several studies have shown that this link between antioxidant activity and the buildup of phenolic compounds under salt stress is true for most plants, and research on the topic is constantly expanding. However, there are specific plant species that have been found to be more resistant to the oxidative damage caused by salt stress, and these are typically plants that have high levels of antioxidant activity. One such plant is rosemary, which has been shown to have high levels of antioxidant activity and resistance to salt stress. This is likely due to the presence of high levels of polyphenols in rosemary, which are powerful antioxidants that can protect cells from damage caused by oxidative stress [46]. Other plants that are known to be resistant to salt stress and have high levels of antioxidant activity are grapefruit, black tea, and green tea.

Flavonoids, polyphenols, tannins, and anthocyanins are some other secondary metabolites that were found in large amounts and may help plants tolerate salt by making their antioxidants work better [47]. Flavonoids are a type of secondary metabolite that is found in large amounts in plants. These compounds can help plants tolerate salt. Polyphenols are another type of secondary metabolite that can help plants tolerates salt. Tannins are types of secondary metabolite that can help plants resist damage from salt. Anthocyanins are another type of secondary metabolite that can help plants resists damage from salt. Yang et al. used saponin as a priming agent to help quinoa plants grow from seeds in salty environments. This is because saponins can get rid of ROS [48]. In this way, by analyzing the changes in metabolites at the whole-metabolome scale and using these metabolic profile changes along with other "omic" analyses such as genome, transcriptome, and proteome analysis, one can figure out the regulatory networks and find biomarkers that control stress responses and can be used to improve plants [39, 49].

2.3 Plant reactions against the salt stress factors

Plants can not get away from potentially dangerous situations, such as those caused by biotic and abiotic stresses, which affect them at every stage of their lives. Stress does not hurt some desert plants, but it can kill others [50]. Some plants have a physiological or biochemical defense against salt stress factors. These plants can accumulate substances that scavenge or chelate ions or that suppress the activity of salt-sensitive enzymes. For example, several plant species accumulate compounds such as salicin or protocatechuic acid that bind to and suppress the activity of ion channels in the cell membrane. Other plants produce compounds such as phenolic acids that inhibit the activity of salt-sensitive enzymes. Plants protect themselves from a wide range of stresses with systems that are complex and well-balanced. There are three main types of systems in plants: photosynthesis, respiration, and homeostasis. Photosynthesis is the process that plants use to create energy from the sun [51]. Respiration is the process that plants use to release energy from the food they eat. Homeostasis is the system that plants use to maintain the correct level of water, minerals, and energy in their cells.

When plants are exposed to high levels of salt, it can disrupt their ability to take up water and nutrients, which can have a negative impact on their metabolism and overall growth. To cope with this stress, plants have evolved a variety of strategies to adjust their metabolism and maintain their physiological functions. Plants adjust their rates of photosynthetic activity, stomatal conductance, transpiration, cell wall architecture, membrane remodeling, cell cycle and division rates, and a variety of other physiological and metabolic activities in response to environmental stresses [52]. This can be achieved by altering the expression of genes that are involved in ion transport and osmotic adjustment. For example, some plants will increase the production of compatible solutes, such as proline and glycine betaine, which help to maintain cellular osmotic balance and protect cellular structures from damage. Other plants may increase the expression of genes involved in ion transport, such as the $Na+/H+$ antiporter, which helps to remove excess sodium from the cell and maintain cellular pH. In addition to these gene expression changes, plants may also alter their metabolic pathways in response to salt stress. For example, some plants may increase their production of antioxidants, which can help to protect against oxidative stress and cell damage. Others may alter their carbohydrate metabolism such as increasing the breakdown of starch to provide energy for growth and maintenance. Stress signals turn on the plant's main metabolism, which makes biosynthetic intermediates for the secondary metabolism. The stress response system and the inducible defense system root stress signals turn on the plant's main metabolism in soil salinity. This increases the rate of uptake of salt ions from the soil, which can lead to increased plant growth and survival in saline soils [53] depend on the ability to turn on or off a number of genes and a number of molecular and cellular processes that have to do with defense. To deal with harsh conditions, plants make SMs from primary metabolites in their cells.

Salt stress causes a reduction in plant growth and development; it also has an effect on carbon combustion, ion uptake, nutritional requirements, and energy metabolism, and it alters the amounts of secondary metabolites, which are crucial physiological markers in salt stress tolerance. Recent advancements have been made in the identification and characterization of the systems that enable plants to resist high salt concentrations and drought stress. These processes allow plants to survive in harsh environments. In plants that are subjected to stressors, such as the presence of a variety of elicitors or signal molecules, the deposition of secondary metabolites frequently takes place [10, 54].

3. Research methods and analysis of metabolic alertness during salt stress

Metabolic alertness during salt stress can be studied using a range of research methods and analyses. These can include *in vitro* studies, animal studies, and human studies. *In vitro* studies involve culturing cells and introducing salt stress to observe the metabolic alertness response. This can be done using a range of techniques such as fluorescence microscopy, flow cytometry, and metabolite assays. Animal studies involve exposing animals to the plants grown under salt stress and observing their metabolic alertness response. This can be done using behavioral tests, metabolic assays, and gene expression analyses. Human studies involve exposing individuals to salt stress and observing their metabolic alertness response. This can be done using various methods such as dietary interventions, metabolic assessments, and physical activity tests. The analysis of metabolic alertness during salt stress will depend on the research method used. For *in vitro* studies, the data can be analyzed using a variety of techniques such as statistical tests, machine learning algorithms, and network analyses. For animal studies, the data can be analyzed using techniques such as behavioral tests, metabolomic profiling, and gene expression analyses. For human studies, the data can be analyzed using techniques such as dietary interventions, metabolic assessments, and physical activity tests [55]. In case of salt stress in plants [56, 57], metabolic alertness during salt stress in plants can be studied by a variety of methods and techniques have been mentioned below:

3.1 Molecular analysis

1. RNA sequencing: RNA sequencing is a powerful tool for studying gene expression in plants under salt stress. It can be used to identify and quantify changes in gene expression in response to salt stress.

2. qPCR: qPCR is a sensitive and quantitative method for measuring gene expression in plants. It can be used to measure the expression of genes involved in metabolic pathways and to identify changes in gene expression in response to salt stress.

3. Proteomics: proteomics is a powerful tool for studying protein expression in plants under salt stress. It can be used to identify and quantify changes in protein expression in response to salt stress.

3.2 Physiological analysis

1. Photosynthesis: photosynthesis is a key metabolic process in plants. It can be used to measure the effects of salt stress on photosynthetic efficiency, as well as to identify changes in photosynthetic rate in response to salt stress.

Research method	Analysis
Observational studies: observing plants in their natural environment to determine their response to salt stress and metabolic alertness.	*Statistical analysis*: analyzing data collected from observational studies and experiments to identify trends in the plant's response to salt stress and metabolic alertness.
Experiments: creating controlled environments to study the effects of different levels of salt stress and metabolic alertness on plants.	*Molecular analysis*: analyzing gene expression, proteome, and metabolome data to identify key genes and pathways involved in the plant's response to salt stress and metabolic alertness.
Molecular techniques: using techniques such as gene expression, proteomics, and metabolomics to study the plant's response to salt stress and metabolic alertness.	*Comparative analysis*: comparing the plants' response to salt stress and metabolic alertness with other species to identify differences and similarities.

Table 1.
The prospective approach to control the salt stress in the crop plants.

2. Enzymatic activity: enzymes are involved in a variety of metabolic pathways in plants. Measurement of enzymatic activity can be used to identify changes in metabolic activity in response to salt stress.

3. ion transport: ion transport is an important process in plants. Measurement of ion transport can be used to identify changes in ion transport in response to salt stress.

3.3 Field experiments

Field experiments can be used to study the effects of salt stress on plant growth and productivity. These experiments can be used to measure changes in plant growth and productivity in response to salt stress, as well as to identify changes in metabolic processes in response to salt stress.

As a whole, the systematic research methodology and its analysis to control the salt stress in the crop plants has resulted in the development of a variety of techniques to maximize crop yields and minimize salt stress (**Table 1**). Firstly, the use of salt tolerance genes and the development of transgenic plants have been found to be effective in controlling salt stress. Secondly, the use of crop rotation, soil amendments, and the application of fertilizers and micronutrients can be helpful in controlling salt stress. Thirdly, the selection of salt-tolerant crop varieties and the use of drip irrigation techniques also play a major role in controlling salt stress. Finally, the use of bioremediation techniques, such as the use of bacteria and fungi to absorb the salt from the soil, can also help in controlling salt stress.

4. Behavior of anthocyanins in soil salinity

Anthocyanins are a type of phenolic pigment that dissolves in water. Pigments are found in glycosylated forms. The red, purple, and blue pigments found in fruits and vegetables are called anthocyanins. The anthocyanin content of berries, currants, grapes, and even some tropical fruits is quite high. Edible vegetables rich in anthocyanins have a red to purple hue and include leafy greens, cereals, roots, and tubers.

There are several different types of anthocyanin pigments, but the most common one is cyanidin-3-glucoside, which is present in a wide variety of plant species [58]. They are glycosides and acylated, while anthocyanidins are 3-hydroxy, 3-deoxy, and O-methylated. Cyanidin, delphinidin, pelargonidin, peonidin, petunidin, and malvidin are some of the most common types of anthocyanidins. Aside from the usual anthocyanins, plants have also been found to have acylated anthocyanins. Acylated anthocyanin has four subtypes: acrylated, coumaroylated, caffeoylated, and malonylated. These compounds become active under the saline stress. Some of them get induced in response to salt stress and hike the amount of synthesis of anthocyanins in the concerned plant. While anthocyanins appear red in acidic conditions, they change to a more typical blue color when the pH level is raised. Extraction, separation, and measurement of anthocyanins have all greatly benefited from the use of chromatography [59]. There have been several studies that link the production of anthocyanins to the induction of various types of stress in plants; however, very few of these studies have examined the rise in anthocyanin levels that occurs in response to salt stress [60].

Anthocyanins in soil salinity can help promote beneficial microbial activity in the soil. Anthocyanins are plant pigments that can help protect the plant against environmental stress such as soil salinity. They can also act as an antioxidant, scavenging for free radicals and reducing oxidative stress. Furthermore, they can induce the production of plant growth regulators and plant hormones, which can stimulate beneficial microbial activity in the soil. This includes bacteria and fungi that can improve soil fertility and nutrient availability. Additionally, anthocyanins can also stimulate the production of enzymes and metabolites that can help to increase the microbial activity in the soil. When plants are subjected to salinity stress, their metabolic processes undergo alterations. These changes manifest themselves first in the vegetative part of the plant (i.e., the leaves) and later in the reproductive organs of the plant (i.e., the flowers). Thus, leaves with intact polyphenols and antioxidants may indicate that the plant's response to salt stress occurred before the 10th day. Salinity may have decreased anthocyanins in flowers because metabolic activities were inhibited, other pigments accumulated, or the plant's antioxidative mechanisms had exhausted their supply [61]. Salinity affects anthocyanin accumulation differently, thus more research is needed. In salinity-exposed tomato genotypes, For instance, Borghesi et al. [62] demonstrated that the accumulation of anthocyanins in two different tomato genotypes reacted in opposing ways when they were exposed to saline. The study suggests that the production and localization of anthocyanins may help the plant acquire resistance to a variety of environmental challenges and that the adaptive advantages of anthocyanins are considerably less muddled in nonreproductive tissues [63].

5. Behavior of flavonoid in soil salinity

In the kingdom of plantae, flavonoids are the secondary metabolites that are present in the widest variety of plant species. These chemicals have a wide variety of physiological and molecular functions in plants, including acting as signaling molecules, contributing to plant defense, influencing the transport of auxin, exhibiting antioxidant activity, and scavenging free radicals [64]. According to Sirin and Aslam [65], among the nonenzymatic antioxidants, phenols and flavonoids make a substantial contribution to removing free radicals in plants, which allows plants to tolerate salt stress by storing the antioxidants in a variety of tissues. Flavonoids are a group of compounds that are found in many plants. These compounds have been studied for their potential health benefits, and many are now known to be beneficial for human health. Flavonoids are a type of polyphenol. They play essential roles in soil salinity regulation [30, 41, 66]. They can reduce the number of ions, such as sodium and chloride, present in the soil solution, which can help maintain soil health and fertility [66]. They have also been shown to promote beneficial microbial activity in the soil, including improving the growth and health of beneficial bacteria and fungi. They also have antioxidant properties, which can help protect plants from environmental stressors such as high levels of salinity. They can also act as signaling molecules, helping plants regulate their response to salt stress. In addition, flavonoids can help promote beneficial microbial activity in the soil, which can improve the soil's fertility and reduce its salt content. This can help boost the soil's fertility, improve water retention and nutrient cycling, and reduce the risk of disease-causing pathogens. They can also help protect plants from environmental stressors such as drought and extreme temperatures [67, 68].

In plant materials, polyphenols can be found in both their free and bound forms. Phenolic acids make up the bulk of the polyphenols that are found in grains and baked goods derived from cereal, and approximately 75% of these are accessible in bound form. Plants' development, reproduction, and eventual grain yield are almost entirely dependent on the leaf proteins [68]. It is not surprising that salt stress causes a decrease in protein content in plant leaves, given that proteins are known to be one of the first targets of reactive oxygen species (ROS) in living organisms. One of the principal sites of ROS damage is the chloroplast, where it leads to the degradation and inactivation of Rubisco, as well as many other changes in the thylakoid and stromal proteins [69].

Because of this, leaf protein concentration is an important indicator of salt stress [70]. One of the most debated topics in the literature is the extent to which genetic diversity affects the protein content of plant leaves when subjected to salt stress. Few studies have looked into how genetics interact with salt stress to affect protein levels. In addition, there is a lot of debate in the scientific literature concerning the way salinity has an effect. For instance, Birhanie et al. [71] found that salt stress similarly reduced total protein concentration in the shoots of two cultivars of wheat (tolerant and sensitive). In contrast, it was discovered that during salt stress, leaf protein concentration varied by genotype and increased [72, 73].

Many studies have been proposed that increasing the accumulation of suitable solutes, such as proline, can improve salt tolerance. Proline, such as other osmolytes, can eventually regulate redox potential through its effects on osmotic adjustment, membrane protection, and enzyme stability in the face of abiotic and biotic stresses [74]. Arabbeigi et al. [75] reported greater proline biosynthesis gene (P5CS) expression in *Ae. cylindrica* may be linked with salt tolerance. It is a bacterial species that is commonly found in salt marshes and coastal habitats. The P5CS gene is known to play a role in proline biosynthesis, which is a process that helps cells to build proteins. The researchers found that *Ae. cylindrica* cells that expressed high levels of the P5CS gene were more resistant to salt stress. Additionally, the study found that deleting the P5CS gene had no impact on the salt tolerance of *Ae. cylindrica* cells. The authors of the study say that the findings suggest that the P5CS gene may play a role in salt tolerance in *Ae. cylindrica*. They say that the findings could help to identify new strategies for preserving salt marsh habitats. In wheat, this finding agrees with that of Kumar et al. [76]. This is congruent to a certain extent with the findings of Ebrahim et al. [77] in barley, who also found that salt stress led to a huge buildup of proline but also found that salt-sensitive genotypes accumulated more proline in their leaves than salt-tolerant ones. Since the osmo-adaptive response includes proline buildup, whether or not it plays a special function in the resistance to abiotic and biotic stresses (such as salinity and drought) is debatable. Several explanations come to mind for this discrepancy. Variations in genotype or species, stress level or duration, and physiological maturity or development are particularly important to note when comparing research.

6. Behavior of phenolic in soil salinity

Increased salinity affects primary carbon metabolism, plant growth and development by osmotic stress, mineral deficiencies, ion toxicity, and physiological and biochemical disturbances. In general, the effect of salinity on plant growth and development is greatly reduced by the presence of phenolic compounds. Phenolic compounds have been shown to reduce the uptake of sodium and other cations, thereby reducing the amount of cations in the soil. Phenolic compounds can also increase the availability of phosphorus, potassium, magnesium, and calcium, which can help to reduce soil salinity. Phenolic compounds can also reduce the amount of bicarbonate and chloride ions in the soil, thereby reducing soil salinity [78].

Phenolic compounds are known to have a strong ability to absorb and react to soil salinity. They are capable of binding and immobilizing ions, including sodium, which helps to reduce the salinity of the soil. These compounds can also reduce the amount of cations present in the soil, which helps to reduce soil salinity [79]. These compounds can also form complexes with other compounds in the soil, preventing them from taking up and dissolving in the soil water, thereby reducing the salinity of the soil.

7. Discussion

Excessive salts can be deposited on the soil surface due to rainfall or irrigation, or salt can be injected into the soil due to oil and gas exploration or mining, both of which lead to salinization. Using salt-based insecticides and fertilizers can also contribute to salinization. Soil salinization is now a major issue in environmental and socioeconomic sustainability around the world. Soil salinization is the process of the soil becoming excessively salty. This can be caused by a number of factors, including overpumping of groundwater, the use of salty irrigation water, and the addition of salt to the soil through fertilizer and other means. The effects of soil salinization are wide-ranging and can have serious consequences for both the environment and human health [9, 80]. Salinization can lead to the formation of salt lakes and ponds, which can damage infrastructure and disrupt water supplies. It can also damage plants and animals and can increase the risk of waterborne diseases. Soil salinization is a growing problem around the world, and it is becoming increasingly important to address it. There are a number of ways to combat soil salinization, and it is important to stay up-to-date on the latest research and developments [81]. Some of the most commonly used methods include (**Table 2**).

Salt lakes are a byproduct of salinization, which can also contribute to the formation of other environmental dangers such as toxic dust storms and the contamination of water supplies and agricultural land. The development of salt-sensitive habitats is one way in which salinization affects human health. Various efforts have been made to address the global problem of salinization. The United Nations Development Programme (UNDP) has launched a number of initiatives to help address the issue, including the Salinization of Soil Information and Capacity Building Centre (SalinSICB), which provides training and resources to help farmers and communities

Irrigation management	This includes adjusting the timing and amount of irrigation, using soil moisture sensors, using different types of water, and improving drainage systems.
Leaching	Leaching is a process by which salts are removed from the soil by flushing them away with water.
Drainage	Proper drainage helps to reduce the number of salts in the soil and can be achieved by installing tile drains or subsurface drains.
Amendments	Adding organic matter to the soil can help to reduce soil salinity and can also improve soil structure and fertility.
Cover crops	Planting cover crops, such as legumes and grasses, can help reduce soil salinity by taking up some of the salts from the soil.
Crop rotation	Rotating different crops in a field can help to reduce soil salinity as different plants have different salt tolerances.
Desalinization	Desalination is a process by which salts are removed from the water and can be used to reduce soil salinity in areas with high levels of water salinity.

Table 2.
Proper management of soil from salinity.

protect their soil from salinization, and the Salinization [80]. The Salinization of Soil Information and Capacity Building Centre (SalinSICB) was launched in March 2008 to support the development of effective soil salinization management practices in countries across the world. The centre provides access to soil salinization information and training to government officials, scientists, farmers, and other practitioners. It also works to build the capacity of countries to manage salinization and address its impacts [76, 82]. Salt stress is a major environmental stressor for plants. It has a negative effect on the environment and is a major contributor to reducing crop production. Plants are unable to grow and develop normally under salt stress, and their levels of secondary metabolites, which serve as important physiological markers of salt stress resistance, are typically altered as a result. Recent successes in identifying and characterizing salt-resistant systems in plants have paved the way for the creation of salt-resistant crops. Here, the chapter discussed the salt stress affects secondary metabolites and several vital plant medicines. During stress, increased cytosolic synthesis of secondary metabolites (anthocyanins, flavonoids, phenolics, and unambiguous phenolic acids) may protect cells from ion-induced oxidative damage by binding the ions and lowering cytoplasmic structural toxicity. Anthocyanins, flavones, phenols, and phenolic acids cause this. Plants adapt to salt stress by changing their metabolism. Understanding these systems may help researchers enhance agricultural output in salt-affected areas.

8. Conclusion and prospects

Over exploitation of groundwater, the use of irrigation water with high salt content, and the application of salt-based pesticides and fertilizers have all contributed to soil salinization, a major problem for environmental and socioeconomic sustainability in many parts of the world. Salinization occurs when salt concentrations in the soil become elevated, leading to a decrease in soil fertility and crop yields. This is often caused by the accumulation of salts in the soil, either due to a lack of leaching or due to the application of salt-based fertilizers and pesticides. The over-extraction of groundwater can also lead to the accumulation of salts in the soil as the water may contain a higher concentration of salts than the soil. In addition, irrigation with water that has a high salt content can lead to the accumulation of salts in the soil. To prevent soil salinization, proper irrigation management, the use of low-salt fertilizers and pesticides, and the proper management of groundwater extraction are all essential. Soil salinization is a problem for both terrestrial and aquatic ecosystems, causing problems for plants, animals, and people. Plants are able to adapt their metabolic processes and physiological activities in order to survive in environments with high levels of salt. This can inhibit plant growth, but plants are able to overcome this obstacle by changing the way they assimilate and use nutrients. When exposed to high levels of salt, plants have evolved mechanisms to adapt their metabolism and preserve physiological processes such as changing the expression of genes involved in ion transport and osmotic adjustment. In addition, plants can upregulate the expression of genes involved in antioxidant enzymes, which help protect the plant from oxidative damage caused by high salinity. Plants also increase the expression of genes involved in the synthesis of compatible solutes, which are osmotically active molecules that increase the osmotic potential of the cell and allow it to retain water. Lastly, plants can reduce their transpiration rates and close their stomata to reduce water loss. It is possible to do this by regulating the expression of genes involved in ion transport and osmotic balance. The expression of genes involved in ion transport and osmotic balance under soil salinity is to introduce a salt stress response gene. A salt stress response gene encodes a transcription factor that responds to salinity stress by upregulating the expression of ion transport and osmotic balance-related genes. This upregulation helps the plant to better cope with stress and improve its overall growth and survival.

There are a number of prospects to control soil salinity using secondary metabolites. Some of these include using antimicrobial agents to control bacterial populations, using plant-derived secondary metabolites to regulate salt uptake, and using plant-derived inhibitors of salinization enzymes. There are several secondary metabolites that have been shown to have the ability to mitigate the effects of soil salinity on plant growth. For example, polyamines such as spermidine and spermine have been found to reduce the toxic effects of salt on plant cells by maintaining ion homeostasis and reducing oxidative stress. Another group of secondary metabolites that have been studied for their potential to control soil salinity are plant growth regulators such as auxins, cytokinins, and gibberellins. These compounds have been found to improve plant growth and productivity under saline conditions by regulating the uptake and distribution of ions, maintaining water balance, and enhancing antioxidant defense mechanisms. In addition to polyamines and plant growth regulators, other secondary metabolites, such as flavonoids, alkaloids, and terpenoids, have also been shown to have potential for controlling soil salinity. These compounds have been found to improve plant growth and productivity by reducing oxidative stress, enhancing nutrient uptake, and regulating ion balance. Soil salinity can have a significant impact on plant metabolism and growth, leading to stunted growth and decreased yields that need to be minimized in the future forecast.

Author details

Maneesh Kumar[1*], Himanshu Bharadwaj[2] and Komal Kumari[2]

1 Department of Biotechnology, Magadh University, Bodh Gaya, Bihar, India

2 Department of Biochemistry, Patna University, Patna, Bihar, India

*Address all correspondence to: kumar.maneesh11@gmail.com

References

[1] Khamidov M, Ishchanov J, Hamidov A, Donmez C, Djumaboev K. Assessment of soil salinity changes under the climate change in the Khorezm region, Uzbekistan. International Journal of Environmental Research and Public Health. 2022;**19**(14):8794. DOI: 10.3390/ijerph19148794

[2] Wicke B, Smeets E, Dornburg V, Vashev B, Gaiser T, Turkenburg W, et al. The global technical and economic potential of bioenergy from salt-affected soils. Energy & Environmental Science. 2011;**4**(8):2669-2681. DOI: 10.1039/C1EE01029H

[3] Hossain MS. Present scenario of global salt affected soils, its management and importance of salinity research. International Research Journal of Biological Sciences. 2019;**1**(1):1-3

[4] Dewi ES, Abdulai I, Bracho-Mujica G, Rötter RP. Salinity constraints for small-scale agriculture and impact on adaptation in North Aceh, Indonesia. Agronomy. 2022;**12**(2):341. DOI: 10.3390/agronomy12020341

[5] Pagare S, Bhatia M, Tripathi N, Pagare S, Bansal YK. Secondary metabolites of plants and their role: Overview. Current Trends in Biotechnology and Pharmacy. 2015;**9**(3):293-304

[6] Ashraf MA, Iqbal M, Rasheed R, Hussain I, Riaz M, Arif MS. Environmental stress and secondary metabolites in plants: An overview. Plant Metabolites and Regulation under Environmental Stress. 2018:153-167. DOI: 10.1016/B978-0-12-812689-9.00008-X

[7] Mainuddin M, Kirby JM. Impact of flood inundation and water management on water and salt balance of the polders and islands in the Ganges delta. Ocean & Coastal Management. 2021;**210**:105740. DOI: 10.1016/j.ocecoaman.2021.105740

[8] Liu JJ, Diao ZH, Xu XR, Xie Q. Effects of dissolved oxygen, salinity, nitrogen and phosphorus on the release of heavy metals from coastal sediments. Science of the Total Environment. 2019;**666**:894-901. DOI: 10.1016/j.scitotenv.2019.02.288

[9] Corwin DL. Climate change impacts on soil salinity in agricultural areas. European Journal of Soil Science. 2021;**72**(2):842-862. DOI: 10.1111/ejss.13010

[10] Gelaye KK, Zehetner F, Loiskandl W, Klik A. Effects of soil texture and groundwater level on leaching of salt from saline fields in Kesem irrigation scheme, Ethiopia. Soil and Water Research. 2019;**14**(4):221-228. DOI: 10.17221/137/2018-SWR

[11] Ewing SA, Sutter B, Owen J, Nishiizumi K, Sharp W, Cliff SS, et al. A threshold in soil formation at Earth's arid–hyperarid transition. Geochimica et Cosmochimica Acta. 2006;**70**(21):5293-5322. DOI: 10.1016/j.gca.2006.08.020

[12] Sytar O, Mbarki S, Zivcak M, Brestic M. The involvement of different secondary metabolites in salinity tolerance of crops. Salinity responses and tolerance in plants. Genome Editing and Systems Biology. 2018;**2**:21-48. DOI: 10.1007/978-3-319-90318-7_2

[13] Irshad A, Rehman RN, Abrar MM, Saeed Q, Sharif R, Hu T. Contribution of rhizobium–legume symbiosis in salt stress tolerance in Medicago truncatula evaluated through photosynthesis, antioxidant enzymes, and compatible

solutes accumulation. Sustainability. 2021;**13**(6):3369. DOI: 10.3390/su13063369

[14] Hao S, Wang Y, Yan Y, Liu Y, Wang J, Chen S. A review on plant responses to salt stress and their mechanisms of salt resistance. Horticulturae. 2021;7(6):132. DOI: 10.3390/horticulturae7060132

[15] Kumar M, Akhtar MS. Application of microbial biotechnology in improving salt stress and crop productivity. Salt Stress, Microbes, and Plant Interactions: Mechanisms and Molecular Approaches. 2019;**2**:133-159

[16] Kiani-Pouya A, Rasouli F, Rabbi B, Falakboland Z, Yong M, Chen ZH, et al. Stomatal traits as a determinant of superior salinity tolerance in wild barley. Journal of Plant Physiology. 2020;**245**:153108. DOI: 10.1016/j.jplph.2019.153108

[17] Chourasia KN, Lal MK, Tiwari RK, Dev D, Kardile HB, Patil VU, et al. Salinity stress in potato: Understanding physiological, biochemical and molecular responses. Life. 2021;**11**(6):545. DOI: 10.3390/life11060545

[18] Parwez R, Aftab T, Gill SS, Naeem M. Abscisic acid signaling and crosstalk with phytohormones in regulation of environmental stress responses. Environmental and Experimental Botany. 2022;**2022**:104885

[19] Kumar M. In silico efficacy of [S]-8-gingerol (a derivative) with 6-gingerol against PT-domain of polyketide synthase A (PksA). IP International Journal of Medical Microbiology and Tropical Diseases. 2021;7:62-64. DOI: 10.18231/j.ijmmtd.2021.014

[20] Gul Z, Tang ZH, Arif M, Ye Z. An insight into abiotic stress and influx tolerance mechanisms in plants to cope in saline environments. Biology. 2022;**11**(4):597. DOI: 10.3390/biology11040597

[21] Kumari A, Kumar M, Sahoo GC, Kumar J. In silico prediction of Glabridin potency against Human Tyrosinase in hyperpigmentation condition. IJRAR-International Journal of Research and Analytical Reviews (IJRAR). 2021;**2021**:2348-2369

[22] Jan R, Asaf S, Numan M, Kim KM. Plant secondary metabolite biosynthesis and transcriptional regulation in response to biotic and abiotic stress conditions. Agronomy. 2021;**11**(5):968. DOI: 10.3390/agronomy11050968

[23] Fiedler D, Clay SA, Westhoff S, Reese CL, Bruggeman SA, Moriles-Miller J, et al. Phytoremediation and high rainfall combine to improve soil and plant health in a North America northern Great Plains saline sodic soil. Journal of Soil and Water Conservation. 2022;**77**(4):381-388. DOI: 10.2489/jswc.2022.00112

[24] Subiantoro R. Assessment of Water Quality for Agricultural Cultivation Irrigation Using the Irrigation Water Quality Index: A Case-Study Land Survey and Evaluation from Kampus Polinela II. IOP Conference Series: Earth and Environmental Science. 2022;**10**(1):012049. DOI: 10.1088/1755-1315/1012/1/012049

[25] Litalien A, Zeeb B. Curing the earth: A review of anthropogenic soil salinization and plant-based strategies for sustainable mitigation. Science of the Total Environment. 2020;**698**:134235. DOI: 10.1016/j.scitotenv.2019.134235

[26] Chen HK, Zeng G, Peng G, Huang M, Shi ZJ et al. Stability, transport and ecosystem effects of graphene in water and soil environments. Nanoscale.

2017;**9**(17):5370-5388. DOI: 10.1039/C6NR09931A

[27] Mashal K, Al-Qinna M, Salahat M, Al-Degs YS, Hamzeh SM. Spatial variations of urban soil salinity and related ions in arid and semiarid areas. Arabian Journal of Geosciences. 2022;**15**(14):1-16. DOI: 10.1007/s12517-022-10540-5

[28] Mishra J, Kumar M, Sharan AK, Kumar J. Detoxification assessment of aflatoxin in Aspergillus flavus under the effect of Calendula officinalis Linn's methanolic extract. Agricultural Science Digest—A Research Journal. 2019;**39**(1):21-25. DOI: 10.18805/ag.D-4874

[29] Jafri H, Khan MSA, Ahmad I. In vitro efficacy of eugenol in inhibiting single and mixed-biofilms of drug-resistant strains of *Candida albicans* and *Streptococcus mutans*. Phytomedicine. 2019;**54**:206-213. DOI: 10.1016/j.phymed.2018.10.005

[30] Fakhri S, Moradi SZ, Farzaei MH, Bishayee A. Modulation of dysregulated cancer metabolism by plant secondary metabolites: A mechanistic review. Seminars in Cancer Biology. 2020;**2020**. DOI: 10.1016/j.semcancer.2020.02.007

[31] Venegas-Molina J, Molina-Hidalgo FJ, Clicque E, Goossens A. Why and how to dig into plant metabolite–protein interactions. Trends in Plant Science. 2021;**26**(5):472-483. DOI: 10.1016/j.tplants.2020.12.008

[32] Salas-González I, Reyt G, Flis P, Custódio V, Gopaulchan D, Bakhoum N, et al. Coordination between microbiota and root endodermis supports plant mineral nutrient homeostasis. Science. 2021;**371**(6525):eabd0695. DOI: 10.1126/science.abd0695

[33] Zhang J, Liu YX, Zhang N, Hu B, Jin T, Xu H, et al. Nrt1.1b is associated with root microbiota composition and nitrogen use in field-grown rice. Nature Biotechnology. 2019;**37**:676-684. DOI: 10.1038/s41587-019-0104-4

[34] Stringlis IA, Yu K, Feussner K, de Jonge R, Van Bentum S, Van Verk MC, et al. Myb72-dependent coumarin exudation shapes root microbiome assembly to promote plant health. Proceedings of the National Academy of Sciences of the United States of America. 2018;**115**:E5213-E5222. DOI: 10.1073/pnas.1722335115

[35] de Vries FT, Griffiths RI, Knight CG, Nicolitch O, Williams A. Harnessing rhizosphere microbiomes for drought-resilient crop production. Science. 2020;**368**:270. DOI: 10.1126/science.aaz5192

[36] Kumar M, Topno RK, Kumar H, Kumar J. Biological conversion of Carcinogenic Aflatoxin B1 to AFB1-8, 9-exo-epoxide and its potential prevention. International Journal of Research and Analytical Reviews. 2018;**5**(4):935-946

[37] Hossain MS, Persicke M, ElSayed AI, Kalinowski J, Dietz K. Metabolite profiling at the cellular and subcellular level reveals metabolites associated with salinity tolerance in sugar beet. Journal of Experimental Botany. 2017;**68**(21-22):5961-5976. DOI: 10.1093/jxb/erx388

[38] Zhu JK. Abiotic stress signaling and responses in plants. Cell. 2016;**167**:313-324. DOI: 10.1016/j.cell.2016.08.029

[39] Sarri E, Termentzi A, Abraham EM, Papadopoulos GK, Baira E, Machera K, et al. Salinity stress alters the secondary metabolic profile of *M. sativa*, *M. arborea* and their hybrid (Alborea). International Journal of Molecular

Sciences. 2021;**22**(9):4882. DOI: 10.3390/ijms22094882

[40] Kumar H, Kumar M, Kumar B, Bharti S, Kumar P, Kumar M, et al. Effect of acetone extract of *Zea mays* L. leaves on the growth of *Aspergillus flavus* and aflatoxin production. International. Journal of Advanced Research. 2017;**5**(8):1667-1671. DOI: 10.21474/IJAR01/5228

[41] Chowdhary V, Alooparampil S, Pandya RV, Tank JG. Physiological function of phenolic compounds in plant defense system. Phenolic compounds-chemistry, synthesis, diversity, non-conventional industrial. Pharmaceutical and Therapeutic Applications. 2021;**2021**. DOI: 10.5772/intechopen.101131

[42] Yang L, Wen KS, Ruan X, Zhao YX, Wei F, Wang Q. Response of plant secondary metabolites to environmental factors. Molecules. 2018;**23**(4):762. DOI: 10.3390/molecules23040762

[43] Lucini L, Borgognone D, Rouphael Y, Cardarelli M, Bernardi J, Colla G. Mild potassium chloride stress alters the mineral composition, hormone network, and phenolic profile in artichoke leaves. Frontiers in Plant Science. 2016;**7**:948. DOI: 10.3389/fpls.2016.00948

[44] Ramalingam A, Kudapa H, PazhamalaLT, WeckwerthW, VarshneyRK. Proteomics and metabolomics: Two emerging areas for legume improvement. Frontiers in Plant Science. 2015;**6**:1116. DOI: 10.3389/fpls.2015.01116

[45] Giuffrida F, Agnello M, Mauro RP, Ferrante A, Leonardi C. Cultivation under salt stress conditions influences postharvest quality and glucosinolates content of fresh-cut cauliflower. Scientia Horticulturae. 2018;**236**:166-174. DOI: 10.1016/j.scienta.2018.03.049

[46] Abdal M, Etemadi N, Nikbakht A, Amirikhah R. The arbuscular mycorrhizal symbiosis alleviating long-term salt stress through the modulation of nutrient elements, osmolytes, and antioxidant capacity in rosemary. Biologia. 2022;**29**:1-8. DOI: 10.1007/s11756-022-01285-3

[47] Ghanem AMF, Mohamed E, Kasem AM, El-Ghamery AA. Differential salt tolerance strategies in three halophytes from the same ecological habitat: Augmentation of antioxidant enzymes and compounds. Plants. 2021;**10**(6):1100. DOI: 10.3390/plants10061100

[48] Yang A, Akhtar S, Iqbal S, Qi Z, Saddiq M, Jacobsen SE. Saponin seed priming improves salt tolerance in quinoa. Journal of Agronomy and Crop Science. 2017:204. DOI: 10.1111/jac.12229

[49] Kumar M, Rana S, Kumar H, Kumar P, Dikhit MR, Mansuri R, et al. Computational, structural and functional aspects of hypothetical protein of Aspergillus flavus pheromone receptor pre-A (PRP-A). Journal of Applied Pharmaceutical Science. 2017;7(7):089-097. DOI: 10.7324/JAPS.2017.70714

[50] Shimizu-Sato S, Tanaka M, Mori H. Auxin–cytokinin interactions in the control of shoot branching. Plant Molecular Biology. 2009;**69**:429. DOI: 10.1007/s11103-008-9416-3

[51] dos Santos TB, Ribas AF, de Souza SG, Budzinski IG, Domingues DS. Physiological responses to drought, salinity, and heat stress in plants: A review. Stress. 2022;**2**(1):113-135

[52] Atkinson NJ, Urwin PE. The interaction of plant biotic and abiotic stresses: From genes to the field. Journal of Experimental Botany.

2012;**63**:3523-3543. DOI: 10.1093/jxb/ers100

[53] Zipfel C, Oldroyd GE. Plant signalling in symbiosis and immunity. Nature. 2017;**543**(7645):328-336. DOI: 10.1038/nature22009

[54] Hussain S, Khalid MF, Hussain M, Ali MA, Nawaz A, Zakir I, et al. Role of micronutrients in salt stress tolerance to plants. In: Plant Nutrients and Abiotic Stress Tolerance. Singapore: Springer; 2018. pp. 363-376. DOI: 10.1007/978-981-10-9044-8_15

[55] Furman D, Campisi J, Verdin E, Carrera-Bastos P, Targ S, Franceschi C, et al. Chronic inflammation in the etiology of disease across the life span. Nature Medicine. 2019;**25**(12):1822-1832

[56] Mellidou I, Ainalidou A, Papadopoulou A, Leontidou K, Genitsaris S, Karagiannis E, et al. Comparative transcriptomics and metabolomics reveal an intricate priming mechanism involved in PGPR-mediated salt tolerance in tomato. Frontiers in Plant Science. 2021;**12**:713984

[57] Kumar A, Behera I, Langthasa M, PrakashNaroju S. Effect of plant growth-promoting rhizobacteria on alleviating salinity stress in plants: A review. Journal of Plant Nutrition. 2022;7:1-26

[58] Khoo HE, Azlan A, Tang ST, Lim SM. Anthocyanidins and anthocyanins: Colored pigments as food, pharmaceutical ingredients, and the potential health benefits. Food & Nutrition Research. 2017;**61**(1):1361779. DOI: 10.1080/16546628.2017.1361779

[59] Mattioli R, Francioso A, Mosca L, Silva P. Anthocyanins: A comprehensive review of their chemical properties and health effects on cardiovascular and neurodegenerative diseases. Molecules. 2020;**25**(17):3809. DOI: 10.3390/molecules25173809

[60] Ben Dkhil B, Denden M. Effect of salt stress on growth, anthocyanins, membrane permeability and chlorophyll fluorescence of okra (*Abelmoschus esculentus* L.) seedlings. American. Journal of Plant Physiology. 2012;7(4):174-183. DOI: 10.3923/ajpp.2012.174.183

[61] Chrysargyris A, Tzionis A, Xylia P, Tzortzakis N. Effects of salinity on tagetes growth, physiology, and shelf life of edible flowers stored in passive modified atmosphere packaging or treated with ethanol. Frontiers in Plant Science. 2018;**9**:1765. DOI: 10.3389/fpls.2018.01765

[62] Borghesi E, González-Miret ML, Escudero-Gilete ML, Malorgio F, Heredia FJ, Meléndez-Martínez AJ. Effects of salinity stress on carotenoids, anthocyanins, and color of diverse tomato genotypes. Journal of Agricultural and Food Chemistry. 2011;**59**:11676-11682. DOI: 10.1021/jf2021623

[63] Eryılmaz F. The relationships between salt stress and anthocyanin content in higher plants. Biotechnology & Biotechnological Equipment. 2006;**20**(1):47-52. DOI: 10.1080/13102818.2006.10817303

[64] Tohidi B, Rahimmalek M, Arzani A. Essential oil composition, total phenolic, flavonoid contents, and antioxidant activity of Thymus species collected from different regions. Iran Food Chemistry. 2017;**220**:153161. DOI: 10.1016/j.foodchem.2016.09.203

[65] Kumar M, Rana S, Kumar H, Kumar J, Mansuri R, Chandra SG. Inhibition of product template (PT) domain of

aflatoxin producing polyketide synthase enzyme of Aspergillus parasiticus. Letters in Drug Design & Discovery. 2017;**14**(7):811-818. DOI: 10.2174/1570180814666170203144914

[66] Kumar S, Pandey AK. Chemistry and biological activities of flavonoids: An overview. The Scientific World Journal. 2013;**2013**. DOI: 10.1155/2013/162750

[67] Manzoor A, Dar IH, Bhat SA, Ahmad S. Flavonoids: Health benefits and their potential use in food systems. Functional Food Products and Sustainable Health. 2020;**2020**:235-256. DOI: 10.1007/978-981-15-4716-4_15

[68] Carpena M, Caleja C, Nuñez-Estevez B, Pereira E, Fraga-Corral M, Reis FS, et al. Flavonoids: A Group of Potential Food Additives with Beneficial Health Effects2021. DOI: 10.1017/jns.2016.41

[69] Sandoval-Ibáñez O, Sharma A, Bykowski M, Borràs-Gas G, Behrendorff JB, Mellor S, et al. Curvature thylakoid 1 proteins modulate prolamellar body morphology and promote organized thylakoid biogenesis in Arabidopsis thaliana. Proceedings of the National Academy of Sciences. 2021;**118**(42):e2113934118. DOI: 10.1073/pnas.2113934118

[70] Kumar M, Kumar H, Topno RK, Kumar J. Analysis of impact of anaerobic condition on the aflatoxin production in Aspergillus parasiticus Speare. Agricultural Science Digest-A Research Journal. 2019;**39**(1):75-78. DOI: 10.18805/ag.A-5161

[71] Birhanie ZM, Yang D, Luan M, Xiao A, Liu L, Zhang C, et al. Salt stress induces changes in physiological characteristics, bioactive constituents, and antioxidants in Kenaf (*Hibiscus cannabinus* L.). Antioxidants. 2022;**11**(10):75. DOI: 10.3390/antiox11102005

[72] Saddiq MS, Iqbal S, Hafeez MB, Ibrahim AM, Raza A, Fatima EM, et al. Effect of salinity stress on physiological changes in winter and spring wheat. Agronomy. 2021;**11**(6):1193. DOI: 10.3390/agronomy11061193

[73] Saddiq MS, Iqbal S, Afzal I, Ibrahim AM, Bakhtavar MA, Hafeez MB, et al. Mitigation of salinity stress in wheat (*Triticum aestivum* L.) seedlings through physiological seed enhancements. Journal of Plant Nutrition. 2019;**42**(10):1192-1204. DOI: 10.1080/01904167.2019.1609509

[74] Ma JF. Role of silicon in enhancing the resistance of plants to biotic and abiotic stresses. Soil Science and Plant Nutrition. 2004;**50**(1):11-18. DOI: 10.1080/00380768.2004.10408447

[75] Arabbeigi M, Arzani A, Majidi MM. Expression profiles of P5CS and DREB2 genes under salt stress in Aegilops cylindrica. Russian Journal of Plant Physiology. 2019;**66**:583-590. DOI: 10.1134/S1021443719040022

[76] Kumar P, Sharma PK. Soil salinity and food security in India. Frontiers in Sustainable Food Systems. 2020;**4**:533781. DOI: 10.3389/fsufs.2020.533781

[77] Ebrahim F, Arzani A, Rahimmalek M, Sun D, Peng J. Salinity tolerance of wild barley Hordeum vulgare ssp. spontaneum. Plant Breeding. 2020;**139**(2):304-316. DOI: 10.1111/pbr.12770

[78] Adusei-Gyamfi J, Ouddane B, Rietveld L, Cornard JP, Criquet J. Natural organic matter-cations complexation and its impact on water treatment: A critical review. Water Research. 2019;**160**:130-147. DOI: 10.1016/j.watres.2019.05.064

[79] DalCorso G, Manara A, Piasentin S, Furini A. Nutrient metal elements in plants. Metallomics. 2014;**6**(10):1770-1788

[80] Mukhopadhyay R, Sarkar B, Jat HS, Sharma PC, Bolan NS. Soil salinity under climate change: Challenges for sustainable agriculture and food security. Journal of Environmental Management. 2021;**280**:111736

[81] Singh A. Soil salinization management for sustainable development: A review. Journal of Environmental Management. 2021;**277**:111383. DOI: 10.1016/j.jenvman.2020.111383

[82] Sharma A, Kumar V, Shahzad B, Ramakrishnan M, Singh Sidhu GP, Bali AS, et al. Photosynthetic response of plants under different abiotic stresses: A review. Journal of Plant Growth Regulation. 2020;**39**:509-531. DOI: 10.1007/s00344-019-10018-x

Chapter 4

Influence of Salinity on *In Vitro* Production of Terpene: A Review

Emine Ayaz Tilkat, Ayşe Hoşer, Veysel Süzerer and Engin Tilkat

Abstract

Terpenes are the largest group of plant secondary metabolites with many biological activities, such as anticancer, antimicrobial, anti-inflammatory, antifungal, and antiviral. They are natural plant products frequently used in many sectors, such as medicine, agriculture, and perfumery. Various biotechnological strategies have been developed to increase terpene production and variety in plants. Among these approaches, using stimulants that induce *in vitro* accumulation of plant secondary metabolites, such as elicitor, is one of the best alternatives. Successful effects of salt (NaCl), an abiotic elicitor, on terpene production in different plant species have been reported. This technique remains relevant as a promising approach to the yet unknown chemistry of many plant species. Therefore, this review aims to appraise the literature available for using NaCl stress as an elicitor in *in vitro* cultures to increase terpene compounds in plants.

Keywords: *in vitro*, salt stress, elicitor, terpene production, NaCl

1. Introduction

Plants are a valuable source of various secondary metabolites used as pharmaceuticals, agrochemicals, flavors, fragrances, colors, biopesticides, and food additives [1]. Plant secondary metabolites (PSM) are low molecular weight organic compounds that do not directly affect plant growth and development but have essential roles as a defensive tool in interacting with the environment and adapting to environmental conditions. Under natural conditions, many PSMs accumulate in different parts of plants (vacuoles, specialized glands, trichomes, and sometimes only at certain developmental stages) to provide functional flexibility under the influence of environmental factors without affecting cellular and physiological developmental pathways [2]. Since these compounds have many properties, such as antioxidant, antimalarial, antifungal, antimicrobial, and antiviral, they are essential in protecting the plant's defense system due to their toxicity. Thus, removing other microbes and herbivores protects them from all kinds of pathogens [3]. PSMs are diverse and numerous chemical compounds derived from primary metabolic pathways by the plant cell. There are over 100,000 known PSMs in the plant kingdom. These compounds are divided into three major classes according to their chemical structures: terpenes, nitrogen-containing compounds (e.g., alkaloids and glucosinolates), and phenolic compounds (e.g., phenylpropanoids and flavonoids) [4].

Terpenes are the largest and most structurally diverse group of natural products, with more than 80,000 characterized compounds [5, 6]. In addition to being a pigment, flavoring and solvent, terpenes have many functions, such as medical thermoprotectant and signal transduction processes. Although there are many types and varieties, it is impressive that different organisms use terpenes for common purposes. Many living organisms, such as microorganisms, fungi, and plants, are protected from abiotic and biotic stresses thanks to their synthesized terpenes [7]. Terpenes are aromatic metabolites found in plants and can improve plants' adaptation to the environment. Some terpenes are of enormous value to humanity due to their application in medicine, industry, and agriculture. To date, 52 antimicrobial terpenes have been identified, including carvacrol, thymol, menthol, geraniol, carnosic acid, quercetin, and allicin [8, 9]. Other terpenes, such as beta-myrcene limonene, pinene, and caryophyllene, may be safe and cost-effective alternatives for treating malaria in the pharmaceutical industry because they have antiplasmodial potential [10, 11]. Recent studies suggest that another terpene, Tanshinone IIA, can prevent the occurrence of atherosclerosis and damage and hypertrophy of the heart [12]. Perillyl alcohol is a monocyclic monoterpene and is of great interest due to its potent antitumor activity.

On the other hand, geraniol has been found to have therapeutic effects on cancer diseases, such as lung, colon, prostate, pancreas, and liver. Artemisinin and its derivatives have been reported to affect tumors significantly and have specific inhibitory effects at low costs [13]. However, terpenes are challenging to produce in large quantities due to their complex chemical structure and low content [14]. Plant production *in vitro* conditions is preferred because it allows the production of plant-specific metabolites using elicitor and precursor compounds and even the increase in the number of metabolites and the synthesis of new metabolites. Elicitors can be defined as a substance that, when delivered to a living cell system in a small concentration, initiates or increases the biosynthesis of specific compounds. Different types of stress used as elicitors may promote or inhibit terpene production [15]. Abiotic elicitors, such as salinity, UV light, temperature, pH, and heavy metals, can stimulate the accumulation of terpenes [16].

Salinity is one environmental factor limiting growth, development, and productivity among the abiotic stress variety. Under salinity conditions, terpenes protect cells from ion-induced oxidative damage [17]. It can also increase the tolerance of biological activity of some plant species, especially *in vitro* conditions. This trend is noteworthy because of the interesting biological properties of terpenes. This approach, which leads to the overproduction of terpenes, is highly desirable, especially in some medicinal plants. Plant production of terpenes against biotic and abiotic stress has been widely studied [18], and in some studies, it has been proven that salinity causes the accumulation of terpenes [16, 19, 20].

Recently, various strategies have been developed to synthesize terpenes, such as optimization of culture media, elicitation, use of precursors, bioreactor cultures, metabolic engineering, immobilization, and biotransformation methods [1]. This section focuses on the effect of salinity on terpene production in medicinal plants grown *in vitro* conditions.

2. Biosynthesis and classifications of plant terpenes

The name "terpene" is derived from the old French word "turpentine" and means "resin" [5]. Plants use two pathways to produce terpenes: the plastidial

2-C-methylD-erythritol-4-phosphate (MEP) pathway and the acetyl-CoA-linked cytosolic mevalonate (MVA) pathway. One of the basic terpene biosynthesis building blocks (C5 isoprene unit) is isopentyl pyrophosphate (IPP), and the other is allylic isomer dimethylallyl pyrophosphate (DMAPP). DMAPP and IPP building blocks combine to form mono-, di-, tri-, tetra-, and polyterpenes with higher molecular weights [21]. Terpenes are volatile unsaturated hydrocarbons with various structural properties [22]. They are commonly found in the leaves, flowers, stems and roots of higher plants, aromatic medicinal plants, citrus, conifers, and eucalyptus species.

The development of chromatographic and spectroscopic methodologies has accelerated the discovery of terpenes and terpenoids [18]. Terpenes and terpenoids are terms that are often used interchangeably. But terpenes undergo oxygenation, hydrogenation, or dehydrogenation to form terpenoids. Terpenes, such as pinene, myrcene, limonene, terpinene, and p-cymene, are compounds with simple hydrocarbon structures. However, terpenoids are a class of modified terpenes with different functional groups and oxidized methyl groups that have been moved or removed [23].

Terpene compounds are classified according to the number of isoprene units they contain. The chemical formula of terpenes is (C5)n. The n, in this formula, indicates the number of isoprenes present in the compound. The number of isoprene units forms the groups that provide the structural diversity of terpenes [14]. Accordingly, terpenes, as shown in **Figure 1**, are hemiterpenes (C5 isoprene), monoterpenes (C10 limonene, menthol, etc.), sesquiterpenes (C15 atractylone, caryophyllene, etc.),

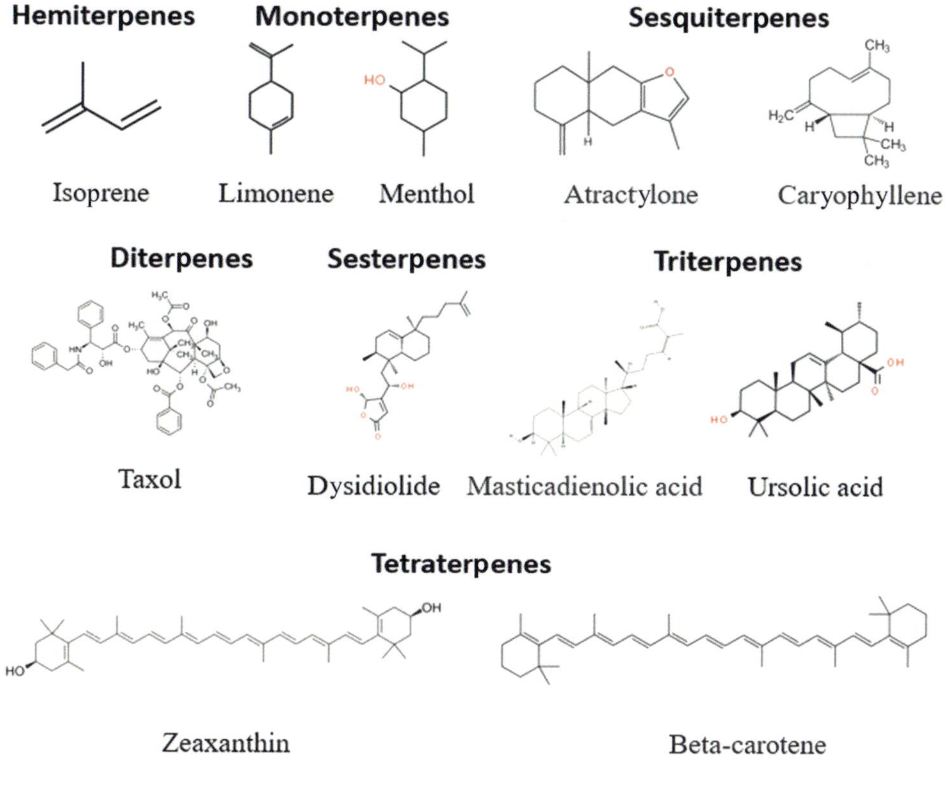

Figure 1.
Classification of terpenes.

diterpenes (C20 taxol, etc.), sesterpenes (C25 disidiolide), triterpenes (C30 masticadienolic acid, ursolic acid, etc.), tetraterpene (C40, zeaxanthin, carotene, etc.) and polyterpenes (C˃40 resin, etc.)

3. Production of terpenes *via* plant cell tissue and organ culture technics

Since terpenes are pharmaceutical compounds with important biological activities, studies on producing these compounds in plants have accelerated in the last 20 years. Plants generally produce low concentrations of terpene in their tissues, with terpene concentration less than 2–3% of a plant's total dry weight. In particular, variation in terpene distribution caused by biotic, abiotic, and seasonal stimuli may vary depending on the plant's chemotype, suggesting that different plants may respond to genetically different stimuli with various terpene syntheses. This situation may differ even between species and individuals [15].

Due to these molecules' complexity and extreme metabolic modifications, their chemical synthesis is inherently tricky, expensive, and relatively low-yield. By 2050, the need for food is expected to double its current level due to overpopulation, and therefore the use of PSMs will increase in the coming decades to meet this demand [24]. With the increase in the world population, modern technologies developed to meet the demand for PSMs and to overcome possible negative situations have begun to be used. Today, many biotechnological techniques such as plant cell and tissue cultures (shoot culture, callus culture, suspension culture, hairy root culture, plant cell immobilization, and bioreactors) and genetic engineering applications are widely used for terpene production [25]. Many other valuable pharmaceutical terpene compounds, such as ginsenosides in Panax ginseng [26, 27], terpenoid indole alkaloids (TIAs) in *Catharanthus roseus* [28, 29], and tanshinones in *Salvia miltiorrhiza* [30, 31], can be produced in high quantities through the shoot, callus, and cell suspension cultures technics. Some studies conducted to increase the production levels of some terpenes and produce them on an industrial scale have focused on bioreactor systems [32]. It has been reported that *Rhizoma zedoariae* cell suspension cultures provide cell proliferation and accumulation of β-element in the bioreactor [33]. Considering that secondary metabolite production may be higher in differentiated tissues, extensive research has also been conducted on hairy root cultures transformed with *Agrobacterium* sp. as an alternative research tool. Extensive research has also been carried out on hairy root cultures transformed with the *Agrobacterium rhizogenes*-mediated transformation method, which reduces the risk of somaclonal variation and provides rapid shoot regeneration, has yielded successful results for the production of terpenes. For example, although the concentration of triterpene detected in the leaves of *Centella asiatica* plants was >2 times higher than in the petiole, the amount of triterpene in the petiole-derived hairy root cultures was 1.4 times higher than in the leaf-derived hairy root cultures. In addition, it was determined that the amount of terpene obtained from leaf and petiole root cultures was higher than that of adventitious roots [34]. Advances in immobilization techniques contribute to a significant increase in the production of high value-added pharmaceutical compounds. *Plumbago rosea*, in which cell cultures were immobilized with 10 mM calcium alginate, plumbagin production was doubled compared to control cells [35].

Furthermore, understanding the function of genes involved in terpene production may lead to discovering new compounds or metabolic pathways that can reveal optimal properties. Accordingly, increased terpene emission was observed in *Nicotiana tabacum* plants' leaves after adding monoterpene synthase genes [36]. In recent years, overproduction, co-overexpression, gene silencing, and genome editing techniques are among the current approaches used for synthesizing different terpene compounds through developments in metabolic engineering.

3.1 Using elicitors for terpene production

It has been shown that certain stress conditions can increase or inhibit terpene production and change the emission pattern and/or amount. Stress conditions generally affect both constitutive and induced terpene emission rates [15]. Treatment of *in vitro* cultures with biotic and abiotic elicitors has been seen as essential to increase the production of desired products [37]. This technique has become the subject of an increasing number of studies today. The elicitor triggers signal transduction and generates secondary signals, stimulating regulatory proteins (transcription factors) that coordinate the expression of biosynthetic genes [38]. Elicitors can be classified as biotic and abiotic elicitors depending on their source and effect on plants (**Figure 2**).

Elicitor type, dose, and application time are the main factors in terpene production [39]. Apart from NaCl, an abiotic elicitor source, there is a lot of literature on methyl jasmonate (MeJA) and salicylic acid (SA). MeJA is a plant growth regulator belonging to the jasmonate family. Extensive research has been conducted on phytohormones, as they can influence many physiological and metabolic processes in plants. Given the great diversity of these compounds with specific biological functions, they are also referred to as biostimulants [24]. Biostimulants are essential as a signaling molecule that mediates intra- and inter-plant communication and modulates plant defense responses, including antioxidant systems. Farag et al. [37] studied the effects of six

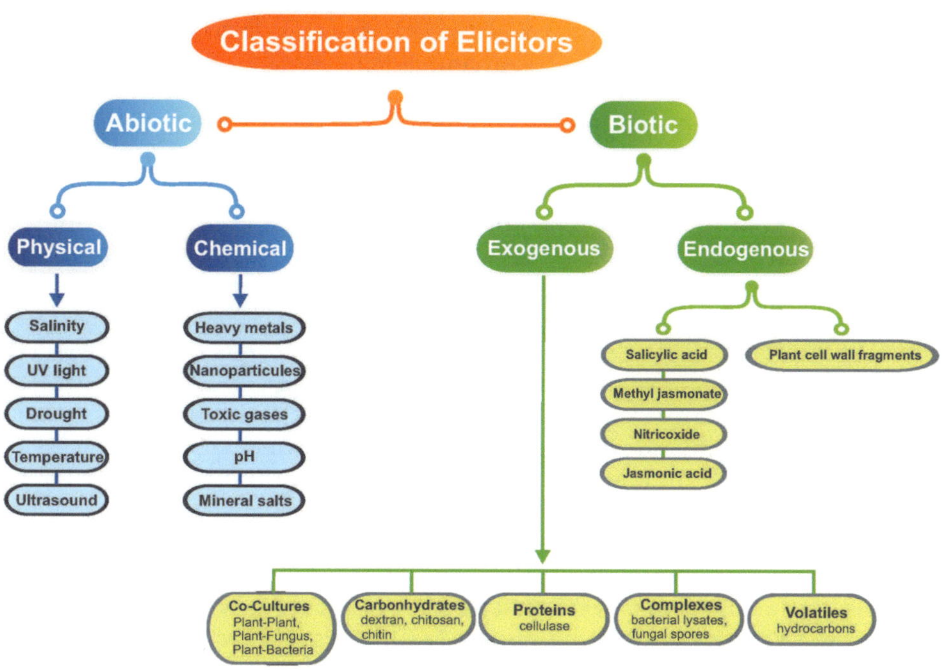

Figure 2.
Classification of elicitors used in secondary metabolite production.

different biotic and abiotic elicitors, including MeJA, SA, $ZnCl_2$, glutathione, and β-glucan (BG; fungal stimulant) wounding, on terpene accumulation in soft coral *Sarcophyton ehrenbergi*. Based on the elicitation process, it has been discovered that the mere inclusion of 0.1 mM SA and 1.0 mM $ZnCl_2$ resulted in a remarkable increase in the levels of sarcophytolide I by 132 and 17 times, respectively, in just 48 hours.

Again, in *Mentha x piperita* plants, where SA or MeJA was applied exogenously, a significant increase in menthol, pulegone, linalool, limonene, and menthone concentrations was achieved with the application of 2.0 mM MeJA [40]. Depending on the dose and time, significant increases in the accumulation of TIAs (eburenin, quebrachamine, fluorocarpamine, pleiocarpamine, tubotaiwine, tetrahydroalstonine, and ajmalicine) occurred in hairy root cultures of *Rhazya stricta*, elicited by the application of different concentrations of MeJA [38]. The roots of *Salvia miltiorrhiza*, a plant rich in tanshinones, an essential active diterpene, are used medicinally for cardiovascular diseases and inflammation. In a study on *Salvia miltiorrhiza*, with 0.2% (v/v) bacterial inoculum as a biotic elicitor, the total tanshinone content of the roots increased more than 12-fold [41]. Li et al. [42] reported that selected genes in the tanshinone biosynthetic pathway of Ag (+), MeJA, and yeast extract were significantly upregulated in *S. castanea* f tomentosa Stib hairy root cultures. Yeast extracts increased the expression level of isopentenyl diphosphate isomerase 13.9-fold at 12 hours. It was determined that yeast extracts increased the expression level of isopentenyl diphosphate isomerase 13.9-fold in 12 h. In contrast, the contents of tanshinone IIA were increased by 1.8-fold and 1.99-fold compared to the control with Ag (+) and MeJA elicitation, respectively.

The addition of the biotic elicitor Anabaena sp. (265 cells/mL) to cell suspension cultures of *Azadirachta indica* triggered the synthesis of the triterpene azadirachtin (0.32 g/μL) [43]. UV-B caused a significant increase in lochnericine concentrations in hairy root cultures of C. roseus. It has also been shown that increasing the exposure time to UV-B up to 20 minutes causes significant increases in lochnericine, serpentine, and ajmalicine and a decrease in horhammericine [44].

3.2 Effect of salt as an elicitor

NaCl is one of the most important abiotic stress factors that cause different changes in plants' morphological, physiological, and biochemical responses, limit their growth and development, and consequently negatively affect total crop production [45, 46]. In the early stages of salinity-induced stress, the ability of roots to absorb water is highly affected and reduced [47]. Higher NaCl concentrations pose a significant threat to the plant by inhibiting physiological processes through osmotic stress, nutrient imbalance, ionic toxicity, and oxidative stress [48]. Oxidative stress is a process in which reactive oxygen species occur as a result of excessive accumulation of sodium (Na^+) and chloride (Cl^-) in plant tissues [49]. After exposure to excess NaCl, plants first sense the potential source of stress and then activate a multifaceted response that includes a signaling network and the synthesis of several compounds that help reduce the effects of high salinity and maintain cellular homeostasis. At this point, secondary metabolites play critical roles in plant adaptation to NaCl stress [4]. Depending on the salinity levels that plants are exposed to *in vitro*, the type and amount of these metabolites they synthesize to survive may vary. Thus, plants can often produce species-specific secondary metabolites in shoots, roots, leaves, etc., at different stages of plant development [10]. Some studies have reported that terpenes exhibit antioxidant activities and thus their function in overcoming oxidative stress [4]. During NaCl stress, terpenes can reduce the consequences of oxidative stress either by reacting directly with intercellular oxidants or altering the signaling of reactive oxygen species. Various terpenes can minimize NaCl stress in different plant species by providing membrane stabilization and direct antioxidant effects. In addition to their antioxidant effects, isoprenes and monoterpenes can react rapidly with ozone and reduce the toxicity caused by NaCl stress. Amphipathic isoprene can prevent membrane and protein degradation by improving hydrophobic interactions between membrane proteins and lipids [20].

Although significant progress has been made on the accumulation mechanisms of terpene compounds in medical plants *via* NaCl elicitation [50], studies are still ongoing on terpene compounds whose chemistry and role are still unknown. To the present day, especially in recent years, research on accumulating these compounds has been presented in **Table 1**. Terpenes and terpenoids are the main components of volatile oils of medical and aromatic plants; it has a variety of chemical compositions ranging from monoterpenes, sesquiterpenes, triterpenes, alcohols, ethers, aldehydes, esters, and ketones [60, 61]. There are 49 different terpenes in the leaves and resins of the male and female mastic tree (*Pistacia lentiscus*) [62]. Tilkat et al. [16] investigated the effects of different NaCl concentrations applied to juvenile shoots of *P. lentiscus in vitro* on the accumulation of anticancer triterpenes, such as ursonic, moronic, oleanolic, masticadienolic, oleanolic, and ursolic acids. Accordingly, while 50 mM NaCl elicitation increased the amount of ursonic acid 2.16 times, 25 mM NaCl elicitation increased it 3.71 times. In addition, masticadienolic acid, which is not found in the control group, was induced *in vitro* leaves *via* elicitated 100 mM NaCl; and ursolic acid was induced *in vitro* leaves and stems by 25, 50, and 100 mM NaCl. It is known that C. roseus, which contains 200 different TIAs, is also a source of vinblastine and vincristine, which have high economic and medical importance. However, since it has been observed that these compounds accumulate in insufficient amounts in the natural growing environment of this plant, it has been tried to increase the production of these compounds with various strategies [63]. Studies have shown that TIA accumulation can be significantly altered through stress mechanisms such as NaCl stress [53, 64] or studies such as overexpression of these genes [65, 66]. In a survey conducted by Fatima et al. [53] using different levels of NaCl concentrations (0, 25, 50, 75, 100, and 125 mM) as an elicitor in various embryogenic tissues of C. roseus grown *in vitro*, the content of vinblastine and vincristine was increased. It was noted that the maximum accumulation of vinblastine and vincristine was obtained by elicitation with 25 mM NaCl. Thymol is a natural volatile monoterpene phenol and the main active ingredient of essential oil obtained from *Thymus vulgaris, Ocimum gratissimum, Carum copticum,* and *Nigella sativa*. Black seed (*N. sativa* L.) is a medicinal plant used to treat many diseases since ancient times. It has been reported that 250 mM NaCl elicitation in *N. sativa* calli leads to high thymol accumulation [19]. Diterpene steviol glycosides obtained from the leaves of the *Stevia rebaudiana* plant are calorie-free natural sweeteners. NaCl and Na_2CO_3 salts were investigated on steviol glycoside production in callus and suspension cultures. In calli, steviol glycosides increased from 0.27% (control) to 1.43% and 1.57%, respectively, with 0.10% NaCl and 0.025% Na_2CO_3. However, in suspension cultures, the same concentrations of NaCl and Na_2CO_3 increased the steviol glycoside content from 1.36 (control) at day 10 to 2.61% and 5.14%, respectively [52]. In addition to the studies on increasing the terpene compounds with a single type of elicitation *in vitro*, more than one type of elicitation has been tried in many plants, and successful results have been obtained. In this context, it has been reported that elicitors, such as NaCl, chitosan, MeJA, SA,

Plant species	Elicitor type	Culture Type	Increased Terpene Compounds	References
Carum copticum	Chitosan + NaCl	Shoot and callus culture	Thymol and p-cymene	[25]
Catharanthus roseus	75 mM NaCl	Callus culture	Vincristine and vinblastine	[51]
Mentha longifolia	4000/ 6000 mg/l for callus, 2000 mg/l for the shoots	Shoot and callus culture	Rosmarinic acid	[51]
Stevia rebaudiana	NaCl and Na_2CO_3	Callus and suspension cultures	Steviol glycosides	[52]
Nigella sativa	250 mM NaCl	Callus culture	Thymol	[19]
Catharanthus roseus	25 mM NaCl	Shoot culture	Vinblastine and vincristine	[53]
Pistacia lentiscus	25 and 50 mM NaCl	Shoot culture	Ursonic acid	[16]
Pistacia lentiscus	100 mM NaCl	Shoot culture	Masticadienolic acid	[16]
Rauwolfia serpentina	100 mM NaCl	hairy root culture	ajmalicine and solasodine	[54]
Datura stramonium	1 and 2 g/l NaCl	Hairy root culture	Hyoscyamine	[55]
Panax ginseng	0.1% NaCl	Hairy root culture	Saponin	[56]
Withania somnifera	50 mM NaCl	Callus culture	Withanolide	[57]
Glycyrrhiza uralensis	0.3%, 0.6%, 1.2%, 1.8%, and 2.4%	Shoot culture	Saponin	[58]
Stevia rebaudiana	150 mM NaCl	Shoot culture	Steviol glycosides	[59]

Table 1.
Effect of salt stress on the accumulation of some terpene compounds in plants.

and jasmonic acid (JA), are widely used in combination or separately for terpene production [25, 67–69]. Razavizadeh et al. (2020) investigated the potential effects of terpene accumulation by examining the combined effect of chitosan and NaCl on *C. copticum* shoot and callus cultures. They reported that combined elicitation led to an increased accumulation of thymol and p-cymene in both shoots and calli [25]. In another study on the production of vincristine and vinblastine from TIAs in C. roseus calluses of both osmotic and NaCl stresses, 75 mM NaCl elicitation significantly increased the content of both bioactive compounds [51]. In another combined elicitation study, the effect of NaCl and phenylalanine on rosmarinic acid production *in vitro* cultures of *Mentha longifolia* was investigated. Five different phenylalanine concentrations (0, 0.5, 5, 10, and 15 mg/l) and four different NaCl concentrations (0, 2000, 4000, and 6000 mg/l) were tested. Shoot tips were more efficient in rosmarinic acid production than callus cultures. In addition, low-concentration NaCl elicitation led to a high accumulation of rosmarinic acid [70].

As discussed above, salt, an abiotic stress factor, actually increases the synthesis of terpenes in plants and is an essential factor for plants to combat the negativities they encounter under natural conditions (**Figure 3**). In addition, it is possible to increase

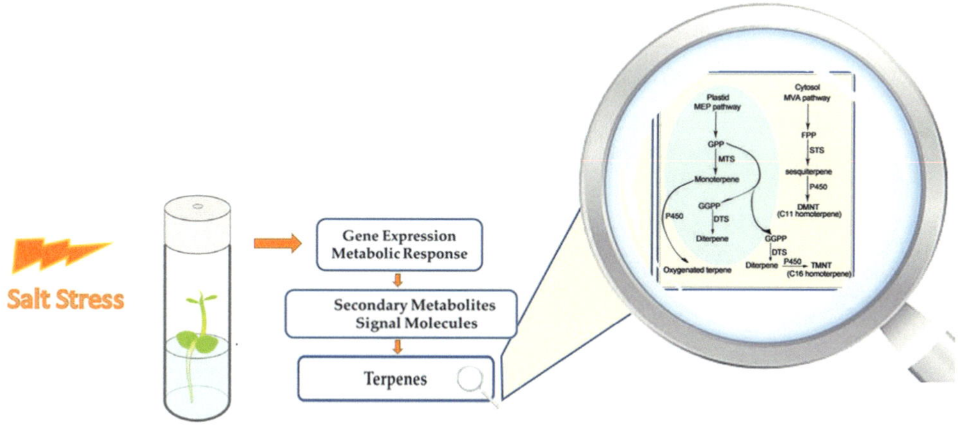

Figure 3.
Terpenes production under NaCl stress [71–73].

the production of terpenes with economic value by plants through salt elicitation carried out *in vitro*. From this point of view, it can be seen that salt, which negatively affects agricultural production, is an essential factor that triggers different metabolic pathways in plants and contributes to the production of many other secondary metabolites, including terpenes, that are beneficial for humanity. Terpenes, in particular, and all secondary metabolites in general, which plants synthesize to make their own life easier and more efficient under salinity, form active compounds used in agricultural and industrial fields and necessary pharmaceutical raw materials for human beings.

4. Future prospects

Many studies have expressed and emphasized elicitation's effect on gene expression in medicinal plants, including salt treatments. With plant tissue culture and genetic engineering applications, regulating the biosynthesis pathways of commercially valuable, potent, effective, and specific terpenes can be modified and improved. By revealing the effects of these applications holistically, it will be possible to produce terpenes, which are especially important in medical terms, on an industrial scale, and at an affordable cost.

Overall, since high salt stress affects the total amount and metabolite profile of secondary metabolites produced, it is clear that salinity regulation can be a promising way to obtain new compounds from plants to produce active PSMs such as raw pharmaceutical materials. In addition, considering the changing climatic conditions and soil profile, it is vital to understand the plant biosynthetic pathways that lead to the production of terpenes, which have a wide range of economic value through salt stress, and ultimately determine how to manipulate these pathways. Accordingly, applying transcriptome and metabolome data to examine gene-metabolite networks at both regulatory and catalytic levels for secondary metabolism in plants will also be an essential approach to reveal the PSM production profile. Additionally, modulation of plant metabolic pathways, leading to target metabolite production through the development of efficient modeling strategies and optimization of growth conditions may help improve the production of important bioactive molecules through bioreactors.

Author details

Emine Ayaz Tilkat[1*], Ayşe Hoşer[1], Veysel Süzerer[2] and Engin Tilkat[1]

1 Faculty of Sciences and Literature, Department of Biology, Batman University, Batman, Türkiye

2 Department of Pharmacy Services, Vocational School of Health Services, Bingöl University, Bingöl, Türkiye

*Address all correspondence to: eayaztilkat@gmail.com

References

[1] Naik PM, Al-Khayri JM. Abiotic and biotic elicitors-role in secondary metabolites production through in vitro culture of medicinal plants. In: Shanker AK, Shanker C, editors. Abiotic and Biotic Stress in Plants-Recent Advances and Future Perspectives. London, UK, London, UK: IntechOpen; 2016. pp. 247-277. DOI: 10.5772/61442

[2] Mohaddab M, El Goumi Y, Gallo M, Montesano D, Zengin G, Bouyahya A, et al. Biotechnology and in vitro culture as an alternative system for secondary metabolite production. Molecules. 2022;**27**:8093-8113. DOI: 10.3390/molecules27228093

[3] Badyal S, Singh H, Yadav AK, Sharma S, Bhushan I. Plant secondary metabolites and their uses. Plant Archives. 2020;**20**:3336-3340 e-ISSN:2581-6063 (online), ISSN:0972-5210

[4] Patel MK, Kumar M, Li W, Luo Y, Burritt DJ, Alkan N, et al. Enhancing salt tolerance of plants: From metabolic reprogramming to exogenous chemical treatments and molecular approaches. Cell. 2020;**9**:2492-2518. DOI: 10.3390/cells9112492

[5] Hussein RA, El-Anssary AA. Plants secondary metabolites: The key drivers of the pharmacological actions of medicinal plants. In: Builders PH, editor. Herbal Medicine. London, UK, London, UK: IntechOpen; 2018. pp. 11-30. DOI: 10.5772/intechopen.76139

[6] Chen L, Pang Y, Luo Y, Cheng X, Lv B, Li C. Separation and purification of plant terpenoids from biotransformation. Engineering in Life Sciences. 2021;**21**:724-738. DOI: 10.1002/elsc.202100014

[7] Cox-Georgian D, Ramadoss N, Dona C, Basu C. Therapeutic and medicinal uses of terpenes. Medicinal Plants. 2019;**12**:333-359. DOI: 10.1007/978-3-030-31269-5_15

[8] Mahizan NA, Yang SK, Moo CL, Song AA, Chong CM, Chong CW, et al. Terpene derivatives as a potential agent against antimicrobial resistance (AMR) pathogens. Molecules. 2019;**24**:2631-2652. DOI: 10.3390/molecules24142631

[9] Gutiérrez-del-Río I, Fernández J, Lombó F. Plant nutraceuticals as antimicrobial agents in food preservation: Terpenoids, polyphenols and thiols. International Journal of Antimicrobial. 2018;**52**(3):309-315. DOI: 10.1016/j.ijantimicag.2018.04.024

[10] Singh E, Khan RJ, Jha RK, Amera GM, Jain M, Singh RP, et al. A comprehensive review on promising antiviral therapeutic candidates identified against main protease from SARS-CoV-2 through various computational methods. JGEB. 2020;**18**:69-81. DOI: 10.1186/s43141-020-00085-z

[11] Shoker RM. A review article: The importance of the major groups of plants secondary metabolism phenols, alkaloids, and terpenes. IJASBT. 2020;**7**:354-358. DOI: 10.31033/ijrasb.7.5.47

[12] Dash DK, Tyagi CK, Sahu AK, Tripathi V. Revisiting the medicinal value of terpenes and terpenoids. In: Meena VS, Parewa HP, Meena SK, editors. Revisiting Plant Biostimulants. London, UK, London, UK: IntechOpen; 2022. DOI: 10.5772/intechopen.102612

[13] Yang W, Chen X, Li Y, Guo S, Wang Z, Yu X. Advances in

pharmacological activities of terpenoids. Natural Product Communications. 2020;**15**:1-13. DOI: 1934578X20903555

[14] Huang Y, Xie FJ, Cao X, Li MY. Research progress in biosynthesis and regulation of plant terpenoids. Biotechnology and Biotechnological Equipment. 2021;**35**:1799-1808. DOI: 10.1080/13102818.2021.2020162

[15] Kopaczyk JM, Warguła J, Jelonek T. The variability of terpenes in conifers under developmental and environmental stimuli. Environmental and Experimental Botany. 2020;**25**:104197. DOI: 10.1016/j.envexpbot.2020.104197

[16] Tilkat E, Ertaş A, Surmuş-Aşan H, Yılmaz MA, Süzerer V. The production of chemical components performing anticancer activity in in vitro culture of shoots, callus and suspension of *Pistacia lentiscus* L.] TÜBİTAK Project (114Z842) Final Report 241p (In Turkish). 2018. https://app.trdizin.gov.tr/publication/project/detail/TVRjMk5UTXo=.

[17] Sytar O, Mbarki S, Zivcak M, Brestic M. The involvement of different secondary metabolites in salinity tolerance of crops. In: Kumar V, Wani SH, Suprasanna P, Tran LSP, editors. Salinity Responses and Tolerance in Plants. Vol. 2. Berlin/Heidelberg, Germany: Springer; 2018. pp. 21-48

[18] Ninkuu V, Zhang L, Yan J, Fu Z, Yang T, Zeng H. Biochemistry of terpenes and recent advances in plant protection. International Journal of Molecular Sciences. 2021;**22**:5710. DOI: 10.3390/ijms22115710

[19] Golkar P, Moradi M, Garousi GA. Elicitation of stevia glycosides using salicylic acid and silver nanoparticles under callus culture. Sugar Technology. 2018;**3**:1-9. DOI: 10.1007/s12355-018-0655-6

[20] Punetha A, Kumar D, Suryavanshi P, Padalia RC. Venkatesha KT environmental abiotic stress and secondary metabolites production in medicinal plants: A review. The Journal of Agricultural Science. 2022;**28**:351-362. DOI: 10.15832/ankutbd.999117

[21] Mani V, Park S, Kim JA, Lee SI, Lee K. Metabolic perturbation and synthetic biology strategies for plant terpenoid production-an updated overview. Plants. 2021;**10**:2179. DOI: 10.3390/plants10102179

[22] Valifard M, Mohsenzadeh S, Kholdebarin B, Rowshan V, Niazi A, Moghadam A. Effect of salt stress on terpenoid biosynthesis in *salvia mirzayanii*: From gene to metabolite. The Journal of Horticultural Science and Biotechnology. 2019;**94**:389-399. DOI: 10.1080/14620316.2018.1505443

[23] Masyita A, Sari RM, Astuti AD, Yasir B, Rumata NR, Emran TB, et al. Terpenes and terpenoids as main bioactive compounds of essential oils, their roles in human health and potential application as natural food preservatives. Food Chemistry: X. 2022;**13**:100217. DOI: 10.1016/j.fochx.2022.100217

[24] Caicedo-López LH, Villagómez-Aranda AL, Sáenz de la OD, Gómez CEZ, Márquez EE, Zepeda HR. Elicitors: Bioethical implications for agriculture and human health. Review Bioet. 2021;**29**:76-86. DOI: 10.1590/1983-80422021291448

[25] Razavizadeh R, Adabavazeh F, Komatsu S. Chitosan effects on the elevation of essential oils and antioxidant activity of *Carum copticum* L. seedlings and callus cultures under in vitro salt stress. Journal of Plant Biochemistry and Biotechnology. 2020;**29**:473-483. DOI: 10.1007/s13562-020-00560-1

[26] Le KC, Jeong CS, Lee H, Paek KY, Park SY. Ginsenoside accumulation profiles in long- and short-term cell suspension and adventitious root cultures in *Panax ginseng*. Horticulture, Environment, and Biotechnology. 2019;**60**:125-134. DOI: 10.1007/s13580-018-0108-x

[27] Qiang B, Miao J, Phillips N, Wei K, Gao Y. Recent advances in the tissue culture of American ginseng (*Panax quinquefolius*). Chemistry & Biodiversity. 2020;**17**:e2000366. DOI: 10.1002/cbdv.202000366

[28] Saiman MZ, Mustafa NR, Pomahočová B, VerbernE M, Verpoorte R, Choi YH, et al. Schulte analysis of metabolites in the terpenoid pathway of *Catharanthus roseus* cell suspensions. PCTOC. 2014;**117**:225239. DOI: 10.1007/s11240-014-0435-2

[29] Panigrahi J, Dholu P, Shah TJ, Gantait S. Silver nitrate-induced in vitro shoot multiplication and precocious flowering in *Catharanthus roseus* (L.) G. Don, a rich source of terpenoid indole alkaloids. PCTOC. 2018;**132**:579-584. DOI: 10.1007/s11240-017-1351-z

[30] Ma XH, Ma Y, Tang JF, He YL, Liu YC, Ma XJ, et al. The biosynthetic pathways of tanshinones and phenolic acids in *salvia miltiorrhiza*. Molecules. 2015;**20**:16235-16254. DOI: 10.3390/molecules200916235

[31] Yu Y, Wang T, Wu Y, Zhou Y, Jiang Y, Zhang L. Effect of elicitors on the metabolites in the suspension cell culture of *salvia miltiorrhiza* Bunge. Physiology and Molecular Biology of Plants. 2019;**25**:229-242. DOI: 10.1007/s12298-018-0605-5

[32] Wawrosch C, Zotchev SB. Production of bioactive plant secondary metabolites through in vitro technologies-status and outlook. Applied Microbiology and Biotechnology. 2021;**105**:6649-6668. DOI: 10.1007/s00253-021-11539-w

[33] Wang GR, Wang H. Cell suspension culture of *Rhizoma zedoariae* in a two-stage perfusion bioreactor system for β-elemene production. In Vitro Cell Development Biology-Plant. 2019;**55**:209-220. DOI: 10.1007/s11627-018-9943-9

[34] Baek S, Han JE, Ho TT, Park SY. Development of hairy root cultures for biomass and triterpenoid production in *Centella asiatica*. Plants. 2022;**2022**(11):148. DOI: 10.3390/plants11020148

[35] Vanisree M, Chen YL, Shu-Fung L, Satish MN, Chien YL, Hsin ST. Studies on the production of some important secondary metabolites from medicinal plants by plant tissue cultures. Botanical Bulletin of Academia Sinica. 2004;**45**:1-22

[36] Trémouillaux-Guiller J. Hairy root culture: An alternative terpenoid expression platform. In: Ramawat K, Mérillon JM, editors. Natural Products. Berlin, Heidelberg: Springer; 2013. pp. 2941-2970. DOI: 10.1007/978-3-642-22144-6_136

[37] Farag MA, Al-Mahdy DA, Meyer A, Westphal H, Wessjohann LA. Metabolomics reveals biotic and abiotic elicitor effects on the soft coral *Sarcophyton ehrenbergi* terpenoid content. Scientific Reports. 2017;**7**:648. DOI: 10.1038/s41598-017-00527-8

[38] Akhgari A, Laakso I, MaaheimoH, ChoiYH, Seppänen-LaaksoT, Oksman-Caldentey KM, et al. Methyljasmonate elicitation increases terpenoid indole alkaloid accumulation in *Rhazya stricta* hairy root cultures. Plants. 2019;**8**:534. DOI: 10.3390/plants8120534

[39] Wang JW, Wu JY. Effective elicitors and process strategies for enhancement of secondary metabolite production in hairy root cultures. Biotechnology Hairy Root System. 2013;**134**:55-89. DOI: 10.1007/10_2013_183

[40] Cappellari LDR, Santoro MV, Schmidt A, Gershenzon J, Banchio E. Improving phenolic total content and monoterpene in *Mentha x piperita* by using salicylic acid or methyl jasmonate combined with rhizobacteria inoculation. International Journal of Molecular Sciences. 2019;**19**:21-50. DOI: 10.3390/ijms21010050

[41] Wu JY, Ng J, Shi M, Wu SJ. Enhanced secondary metabolite (tanshinone) production of *salvia miltiorrhiza* hairy roots in a novel root-bacteria co-culture process. Applied Microbiology and Biotechnology. 2007;**77**:543-550. DOI: 10.1007/s00253-007-1192-5

[42] Li B, Wang B, Li H, Peng L, Ru M, Liang Z, et al. Establishment of *Salvia castanea* Diels f. tomentosa Stib. hairy root cultures and the promotion of tanshinone accumulation and gene expression with Ag^+, methyl jasmonate, and yeast extract elicitation. Protoplasma. 2016;**253**:87-100. DOI: 10.1007/s00709-015-0790-9

[43] Poornasri Devi B, Vimala A, Sai I, Chandra S. Effect of cyanobacterial elicitor on neem cell suspension cultures. Indian Journal of Science and Technology. 2008;**1**:1-5. DOI: 10.17485/ijst/2008/v1i7.14

[44] Binder BY, Peebles CA, Shanks JV, San KY. The effects of UV-B stress on the production of terpenoid indole alkaloids in *Catharanthus roseus* hairy roots. Biotechnology Progress. 2009;**25**:861-865. DOI: 10.1002/btpr.97

[45] Kuşvuran Ş, Havadar S, Daşgan HY. Domateste (*Solanum lycopersicum* L.) Dışsal IAA Uygulamalarının Tuza Tolerans Üzerindeki Etkisi. Uluslararası Tarım ve Yaban Hayatı Bilimleri Dergisi. 2022;**8**:25-37. DOI: 10.24180/ijaws.1033635

[46] Shelake RM, Kadam US, Kumar R, Pramanik D, Singh AK, Kim JY. Engineering drought and salinity tolerance traits in crops through CRISPR-mediated genome editing: Targets, tools, challenges, and perspectives. Plant Communications. 2022;**3**:100417. DOI: 10.1016/j.xplc.2022.100417

[47] Xu C, Tang X, Shao H, Wang H. Salinity tolerance mechanism of economic halophytes from physiological to molecular hierarchy for improving food quality. Current Genomics. 2016;**17**:207-214. DOI: 10.2174/13892 02917 66616 02022 15548

[48] Shrivastava P, Kumar R. Soil salinity: A serious environmental issue and plant growth promoting bacteria as one of the tools for its alleviation. Saudi Journal of Biology and Science. 2015;**22**:123-131. DOI: 10.1016/j.sjbs.2014.12.001

[49] Farouk S, AL-Huqail AA. Sustainable biochar and/or melatonin improve salinity tolerance in borage plants by modulating osmotic adjustment, antioxidants, and ion homeostasis. Plants. 2022;**11**:765. DOI: 10.3390/plants11060765

[50] Isah T, Umar S, Mujib A, Sharma MP, Rajasekharan PE, Zafar N, et al. Secondary metabolism of pharmaceuticals in the plant in vitro cultures: Strategies, approaches, and limitations to achieving higher yield. PCTOC. 2018;**132**:239-265. DOI: 10.1007/s11240-017-1332-2

[51] Alaakel SN. Improving the production of indole alkaloids from Madagascar periwinkle (*Catharanthus roseus* L.) using stressful tissue cultures,

Damascus university. Faculty of Agricultural Engineering. 2021;**1**:184

[52] Gupta P, Sharma S, Saxena S. Effect of salts (NaCl and Na_2CO_3) on callus and suspension culture of *Stevia rebaudiana* for steviol glycoside production. Applied Biochemistry and Biotechnology. 2014;**172**:2894-2906. DOI: 10.1007/s12010-014-0736-2

[53] Fatima S, Mujib A, Dipti T. NaCl amendment improves vinblastine and vincristine synthesis in *Catharanthus roseus*: A case of stress signalling as evidenced by antioxidant enzymes activities. PCTOC. 2015;**121**:445-458. DOI: 10.1007/s11240-015-0715-5

[54] Srivastava M, Sharma S, Misra P. Elicitation based enhancement of secondary metabolites in *Rauwolfia serpentina* and *Solanum khasianum* hairy root cultures. Pharmacognosy Magazine. 2016;**12**(Suppl 3):S315-S320. DOI: 10.4103/0973-1296

[55] Khelifi L, Zarouri B, Amdoun R, Harfi B, Morsli A, Khelifi-Slaoui M. Effects of elicitation and permeabilization on hyoscyamine content in *Datura stramonium* hairy roots. Advances in Environmental Biology. 2011;**5**:329-334

[56] Jeong GT, Park DH. Mass production of transformed hairy root for secondary metabolites: A case study of *Panax ginseng* hairy roots. In: Malik S, editor. Production of Plant Derived Natural Compounds through Hairy Root Culture. Cham: Springer; 2017. pp. 183-201. DOI: 10.1007/978-3-319-69769-7_10

[57] Sabir F, Sangwan RS, Kumar R, Sangwan NS. Salt stress-induced responses in growth and metabolism in callus cultures and differentiating ın vitro shoots of ındian ginseng (*Withania somnifera* Dunal). Journal of Plant Growth Regulation. 2012;**31**:37-548. DOI: 10.1007/s00344-012-9264-x

[58] Nishidono Y, Niwa K, Tanaka K. Effect of salt stress on the accumulation of triterpenoid saponins in aseptic cultured *Glycyrrhiza uralensis*. Plant Growth Regulation. 2023;**100**(1):25-31. DOI: 10.1007/s10725-022-00933-7

[59] Gerami M, Majidian P, Ghorbanpour A, Alipour Z. *Stevia rebaudiana* Bertoni responses to salt stress and chitosan elicitor. Physiology and Molecular Biology of Plants. 2020;**26**:965-974

[60] Spyridopoulou K, Tiptiri-Kourpeti A, Lampri E, Fitsiou E, Vasileiadis S, Vamvakias M, et al. Dietary mastic oil extracted from *Pistacia lentiscus* var. chia suppresses tumour growth in experimental colon cancer models. Scientific Reports. 2017;**19**(7):3782. DOI: 10.1038/s41598-017-03971-8

[61] Noriega P. Terpenes in essential oils: Bioactivity and applications. In: Perveen S, Al-Taweel AM, editors. Terpenes and Terpenoids. London, UK, London, UK: IntechOpen; 2020. pp. 1-13. DOI: 10.5772/intechopen.93792

[62] Onay A, Yıldırım H, Uncuoğlu AA, Özden Çiftçi Y, Tilkat E. Sakız ağacı (*Pistacia lentiscus* L.) yetiştiriciliği. Dicle Üniversitesi Basımevi. 2016;**1**:113

[63] Liu Y, Patra B, Singh SK, Paul P, Zhou Y, Li Y, et al. Terpenoid indole alkaloid biosynthesis in *Catharanthus roseus*: Effects and prospects of environmental factors in metabolic engineering. Biotechnology Letters. 2021;**43**:2085-2103. DOI: 10.1007/s10529-021-03179-x

[64] Ramani S, Jayabaskaran C. Enhanced catharanthine and vindoline production

in suspension cultures of *Catharanthus roseus* by ultraviolet-B light. Journal of Molecular Signaling. 2008;**3**:9. DOI: 10.1186/1750-2187-3-9

[65] Mehrotra S, Srivastava V, Rahman L, Kukreja A. Overexpression of a *Catharanthus tryptophan* decarboxylase (tdc) gene leads to enhanced terpenoid indole alkaloid (TIA) production in transgenic hairy root lines of *Rauwolfia serpentina*. PCTOC. 2013;**115**:377-384. DOI: 10.1007/s11240-013-0369-0

[66] Yamada Y, Sato F. Transcription factors in alkaloid engineering. Biomolecules. 2021;**11**:1719. DOI: 10.3390/biom11111719

[67] Sharmila R, Subburathinam KM. Effect of signal compounds on andrographolide in the hairy root cultures of *Andographis paniculata*. International Journal of Pharm Science. 2013;**4**:1773. DOI: 10.13040/IJPSR.0975-8232.4(5).1773-76

[68] Zaheer M, Giri CC. Enhanced diterpene lactone (andrographolide) production from elicited adventitious root cultures of *Andrographis paniculata*. Research on Chemical Intermediates. 2017;**43**:2433-2444. DOI: 10.1007/s11164-016-2771-9

[69] Alsoufi ASM, Pączkowski C, Szakiel A, Długosz M. Effect of jasmonic acid and chitosan on triterpenoid production in *Calendula officinalis* hairy root cultures. Phytochemistry Letters. 2019;**31**:5-11. DOI: 10.1016/j.phytol.2019.02.030

[70] El-shennawy OA, El-torky MG, El-mokadem HE, Abass BI. Effect of NaCl and phenylalanine on the production of some secondary metabolites in in vitro cultures of *Mentha longifolia*. Alexandria Science Exchange Journal. 2017;**38**:577-587

[71] Zhou HC, Shamala LF, Yi XK, Yan Z, Wei S. Analysis of terpene synthase family genes in *Camellia sinensis* with an emphasis on abiotic stress conditions. Scientific Reports. 2020;**10**:933. DOI: 10.1038/s41598-020-57805-1

[72] Vergine M, Nicolì F, Sabella E, Aprile A, De Bellis L, Luvisi A. Secondary metabolites in *Xylella fastidiosa*–plant ınteraction. Pathogens. 2020;**9**:675. DOI: 10.3390/pathogens9090675

[73] Wang MR, Bettoni JC, Zhang AL, Lu X, Zhang D, Wang QC. In vitro micrografting of horticultural plants: Method development and the use for micropropagation. Horticulturae. 2022;**8**:576. DOI: 10.3390/horticulturae8070576

Chapter 5

Phytohormones-Assisted Management of Salinity Impacts in Plants

Naser A. Anjum, Asim Masood, Faisal Rasheed, Palaniswamy Thangavel and Nafees A. Khan

Abstract

The salinity of soils has been significantly limiting crop production in most arid and semi-arid regions of the world. Plant hormones (phytohormones), small molecules with versatile roles in plants can be a sustainable approach for minimizing the major salinity-impacts in plants. Most phytohormones are reported to regulate various signaling cascades interrelated with plant development and stress-resilience and -coping mechanisms. In addition to regulating photosynthesis and related variables, phytohormones also modulate nutrient homeostasis, source-sink capacity, osmoregulation, and antioxidant defense systems in plants under abiotic stresses including soil salinity. Molecular studies have confirmed the coordination between phytohormones and signaling networks, which in turn also maintains ionic homeostasis and plant-salinity tolerance. This chapter aims to appraise the literature available on the role of 10 well-characterized stress response hormones (abscisic acid, ABA; ethylene; salicylic acid, SA; jasmonic acid, JA; and nitric oxide, NO) and also other growth-promoting hormones (such as auxins, gibberellins, GA; cytokinins, CKs; brassinosteroids, BRs; and strigolactones, SLs) in the management of salinity impacts in plants. The discussion outcomes may help in devising and furthering the strategies aimed at sustainably strengthening plant-salinity tolerance.

Keywords: abiotic stress, soil salinity, plant health, phytohormones, stress tolerance

1. Introduction

Human's major three requirements, namely food, clothing, and shelter are mainly provided by agricultural activity. Unfortunately, the health and productivity of most agricultural crops are impacted by both biotic (insect pests and disease pathogens) and abiotic (temperature, drought, flooding, salinity, heavy metals, radiation, nutrient deficiency, and excess) stress factors. In turn, these stress factors, in isolation and/or combination, are significantly threatening global food security. In particular, abiotic stresses are responsible for about 51–82% of annual loss in yield of the major food crops in the world agriculture. Interestingly, abiotic stress impacts on agricultural crops and food security are further aggravated by the deteriorating agro-climatic conditions [1–4]. According to the United Nations Population Division, the world population will reach 9 billion in 2037 and 10 billion in 2058 [5]. It would be far easier to feed the projected human population because most grain crops exhibit about 1% annual yield (much < world's population growth rate), and direct feeding of people shares only 55% of the world's crop calories [6, 7]. Among the major environmental challenges, the salinization of soils (soil salinity) has been inducing most land degradation, constituting a primary limit on crop health and productivity, and thus, is threatening agriculture across the world [8].

2. Soil salinity: causes, status, and major impacts

2.1 Causes and status

Soils exhibiting the saturation paste extracts (ECe) in the root zone as electrical conductivity (EC) > 4.0 dS m^{-1} (≈40 mM NaCl) at 25°C and 15% exchangeable Na^+ ion are considered saline. There can be five major classes of soil salinity: non-saline (EC = 0–2 dS m^{-1}; low saline (EC = 2–4 dS m^{-1}; moderately saline (EC = 4–8 dS m^{-1}; highly saline (EC = 8–16 dS m^{-1}; and extremely saline (EC = ≥ 16 dS m^{-1}) (**Figure 1**) [9–12]. Most saline soils exhibit Na^+ as an anion and Cl^- as a cation [13]. However, Na^+, Ca^{2+}, and Mg^{2+} are the major cations component of total soluble salts in soils, whereas Cl^-, SO_4^{2-}, and carbonates (CO_3^{2-} and HCO_3^-) are the major anions in total soluble salts in soils. Despite the aforementioned fact, most studies aimed at exploring plant-salinity responses and tolerance have considered Na^+ and Cl^-; and have largely ignored other cations (Ca^{2+} and Mg^{2+}) and/or anions (SO_4^{2-}, CO_3^{2-}, and HCO_3^-) [14]. Both Na^+ and Cl^- ions are the most widespread causes of soil salinity, where Cl^- is more dangerous than Na^+ since it is responsible for many physiological disorders in plants [15].

Notably, the soil salinity may be developed as natural (geological, hydrological, and pedological processes) or induced by human activity (human-caused factors). The long-term natural accumulation of salts (including Cl^- of Na^+, Ca^{2+}, and Mg^{2+}

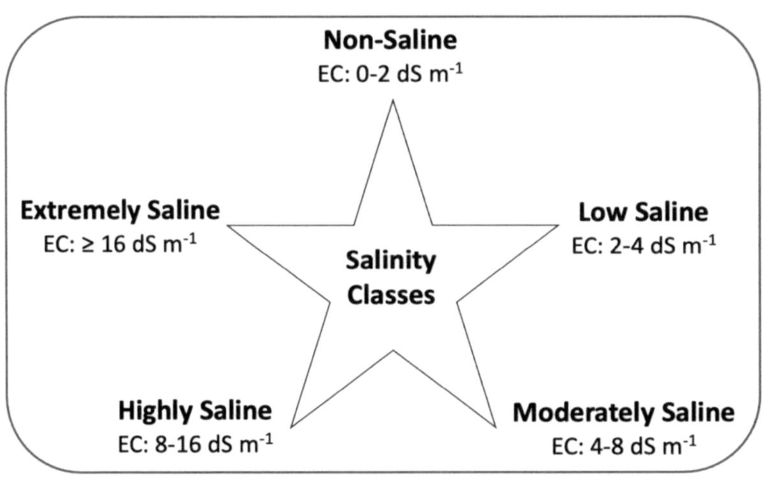

Figure 1.
Schematic representation of the major classes of soil salinity [9, 10]. dSm^{-1}, decisiemens per meter.

and sometimes SO_4^{2-} and CO_3^{2-}) in the soil or surface water contributes to the primary or natural salinity. On the other hand, the disruption of the hydrologic balance of the soil between water applied (irrigation or rainfall) and water used by crops (transpiration), as a result of anthropogenic activities, causes secondary soil salinity [16]. In fact, the secondary salinization (rising NaCl levels in groundwater) impacts irrigated land and eventually leads to the loss of agricultural soils [17]. About 1125 million hectares worldwide have already been impacted by the soil salinization [18]. Huge annual global loss in crop production (≈ US$27.3 billion) has been reported in saline soils in irrigated areas, representing about 20% of the total salinity-affected soils, mainly in North America, Oceania, and the Middle East. Interestingly, salinity may impact half of the world's irrigated land by 2050 [19–22].

2.2 Major impacts

In terms of the tolerance to salinity levels, plants are grouped into two categories, namely halophytes (salinity-tolerant and adapted to salinized environments) and glycophytes (salinity-sensitive plants). Unfortunately, most agricultural crop plants are glycophytes (salt-sensitive), where soil salinization adversely inhibits growth, metabolism, development, and productivity (yield). Thus, the salinity-sensitivity of most crop plants, increasing rate of land salinization, and salinity-caused serious loss in crop productivity are challenging food security. Soil salinity significantly limits the growth, metabolism, seed germination, flowering, fruiting, and productivity (crop yields) in most crop plants, mainly as a result of osmotic stress and ionic stress [3, 23].

2.2.1 Osmotic stress and ionic stress

High soil salinity is bound to cause osmotic stress, followed by ionic stress. These two high soil salinity-caused direct stresses (ionic and osmotic) induce secondary stresses such as oxidative stress. Increased Na^+ ions in the soil mainly cause osmotic stress. Equilibrium in ion homeostasis is severely disturbed, which includes an increase in the Na^+ levels, K^+ efflux, K^+ leakage, K^+ deficiency in the cytosol, replacement of Ca^{2+} with Na^+, and eventual impaired Na^+/K^+ ratio, nutritional disbalance, and disrupted enzyme activity [11, 12, 15]. Significantly reduced water absorption, decreased osmotic potential, closure of the stomata, inhibited CO_2-influx, and impaired downstream processes were reported as a result of osmotic stress in plants [22–24].

2.2.2 Oxidative stress and antioxidant metabolism

Elevated salinity-induced ionic and osmotic stresses are bound to cause oxidative stress, which is a physiological condition of elevated cellular generation of reactive oxygen species (ROS) and or diminished scavenging (antioxidant-mediated) of ROS leading to impaired cellular redox homeostasis. In turn, elevation in the generation of ROS and impaired ROS metabolism/scavenging has been widely reported in plants under soil salinization. The list of major ROS includes $O_2^{\bullet-}$, •OH, H_2O_2, and $_1O^2$ are produced in different including chloroplast, mitochondria, apoplast, and peroxisome, under normal conditions of stress such as salinity. If not metabolized and/or scavenged, most elevated cellular ROS may lead to cytotoxicity and severe damage to biological molecules such as proteins, lipids, and nucleic acids. In order to avoid the harmful effects of oxidative stress, plants have triggered an antioxidant defense system based on two types of machinery [12, 25]. Fortunately, plants are endowed with an antioxidant defense system comprising (enzymes and non-enzymes/ antioxidant metabolites), which in isolation and/or combination tend to scavenge/ metabolize most ROS and thereby avert elevated ROS-caused consequences [26, 27]. Major enzymatic antioxidants comprise superoxide dismutase (SOD), catalase (CAT), guaiacol peroxidase (GPX), glutathione sulfotransferase (GST), ascorbate peroxidase (APX), monodehydroascorbate reductase (MDHAR), dehydroascorbate reductase (DHAR) and glutathione reductase (GR). On the other hand, ascorbate (AsA), glutathione (GSH), carotenoids, tocopherols, and phenolics are among the major non-enzymatic antioxidants in the plant stress defense system [27–34].

Given the rising demands on crop yield for keeping up with the burgeoning human population, strategies for sustainably maintaining an optimum crop plant health under increasing soil salinization and climate change and the development of salt-tolerant crop varieties are being exhaustively explored.

3. Phytohormones and plant-salinity tolerance

Phytohormones stand second to none among the regulatory compounds and chemical messengers in terms of their importance in almost every aspect of plant life. Although phytohormones are small molecules, they are widely known to play key roles in plant growth, development, and various plant physiological processes. These also regulate internal and external stimuli, stress-involved signal transduction pathways, oxidative stress-scavenging, and stress tolerance in plants [35–37]. Plants under salinity stress usually tend to accumulate a range of osmolytes (osmoprotectants/compatible solutes), namely proline, glycine betaine, polyamines, and sugars. Interestingly, the cellular levels of most of these osmolytes are significantly modulated by varied phytohormones [38].

Employing the protocols, based mainly on the use of various phytohormones, may help sustainably improve optimum growth, metabolism, photosynthesis and productivity (yield) under salinity-affected soils, thereby minimizing increasing strain on global food security. Notably, the list of well-characterized stress response hormones includes abscisic acid (ABA), ethylene, salicylic acid (SA), and jasmonic acid (JA). In contrast, phytohormones classified as growth promotion hormones are auxin, gibberellin (GA), cytokinins (CKs), brassinosteroids (BRs), nitric oxide (NO), and strigolactones (SLs) [39, 40].

Apart from presenting a brief overview, the following sections attempt to enlighten the major roles (and the basic mechanisms involved) of ABA, auxins, BRs, CKs, ethylene, GAs, JA, NO, SA, and SLs in plant-salinity tolerance (**Figure 2**).

3.1 Abscisic acid

Chemically, a sesquiterpenoid, 15-C compound, abscisic acid (ABA) is a naturally occurring and enigmatic stress phytohormone widely known to play key roles in plant growth and development. Important processes including leaf abscission, seed dormancy, embryo morphogenesis, stomatal opening and cell turgor maintenance are known to varyingly involve ABA [37, 41–43]. ABA has also been argued as a modulator of plant's adaptive stress responses via integrating various stress signals, controlling downstream responses, biosynthesizing dehydrins, osmolytes, and protective proteins, and regulating protein-encoding genes [43–45]. ABA-supply

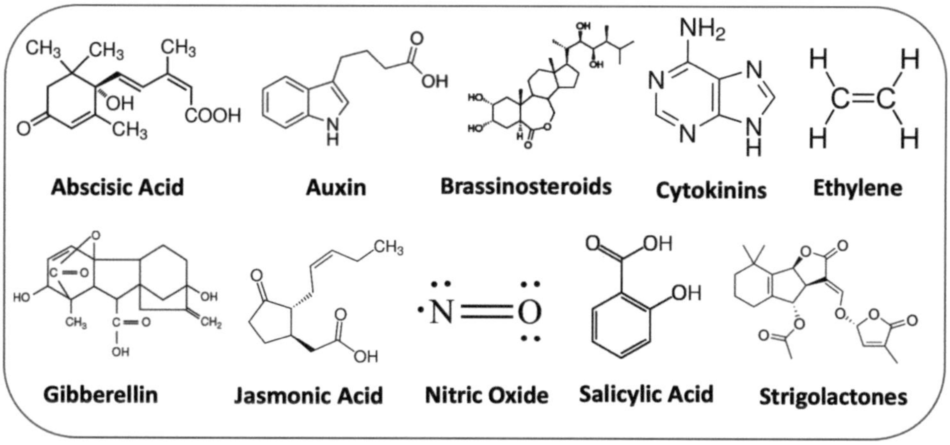

Figure 2.
Schematic representation of the typical structures of phytohormones discussed in the chapter.

helped pepper (*Capsicum annuum*) seeds to exhibit high germination percentage, radicle emergence, and cotyledon expansion of seeds under NaCl stress mainly as a result of low expression in seeds of ABA signaling components such as *CaABI*, *CaPYL2*, *CaPYL4*, *CaSnRK2.3*, and *CaSnRK2.6* [46]. ABA-nitrogen coordination alleviated salinity-inhibited photosynthetic potential in mustard (*Brassica juncea*) by improving proline accumulation and antioxidant activity [47]. Major physiological mechanisms underlying ABA-induced salinity tolerance in plants may also involve enhanced activity of antioxidant enzymes (CAT, APX, peroxidase, and POD) and the contents of antioxidant non-enzyme/metabolites (AsA and GSH) [48]. Additionally, significant reductions in Na^+ content, increased contents of K^+, Mg^{2+}, and Ca^{2+}; and that of hormones such as 1-aminocyclopropane carboxylic acid, trans-zeatin, N6-isopentenyladenosine, indole-3-acetic acid (IAA), and ABA were also observed in ABA-supplied plants under salinity stress [48]. Involvement of ABA signaling in reduction of transpiration flow, regulation of Na^+ ion homeostasis and antioxidant enzyme activities was reported to induce salinity tolerance in wheat (*Triticum aestivum*) seedlings [49]. Earlier, ABA-mediated improved Indica rice (*Oryza sativa*)-tolerance to salinity stress involved the calmodulin signaling cascade and the ABA-mediated induction of *OsP5CR* gene expression in osmolyte (proline) accumulation [50].

3.2 Auxins

A chemical messenger involved in the light and gravity-stimulated shoot-to-root transport of a 'growth stimulus' was argued to be auxin, whose chemical nature was later identified as IAA [51, 52]. Chemically similar to the amino acid tryptophan, IAA is the representative and most studied auxin in plants [51–54]. Auxins are mainly involved in cell division, cell elongation, and cell differentiation [45]. However, auxins can also regulate plant abiotic stress responses, where its homeostasis, distribution, and metabolism can be modulated by most abiotic stress factors [35, 45, 52]. Auxins can mediate the root growth plasticity in response to salinity stress [55, 56]. Earlier, a significant remodeling of root architecture was reported under high salinity, which was argued due to salinity-led altered auxin-accumulation and -redistribution

[57, 58]. Pre- or post-treatment of seeds with IAA significantly alleviated salinity impacts and improved seed germination and early seedling establishments of *T. aestivum* under salinity stress [59]. Hence, an optimum concentration and timely exogenous application of auxins would be a promising approach for countering the salinity stress impacts in crop plants [36].

3.3 Brassinosteroids

Considered ubiquitous in the plant kingdom, polyhydroxy steroidal phytohormones, namely brassinosteroids (BRs), promote growth, seed germination, rhizogenesis, and senescence in plants, as well as their stress-tolerance capacity. Notably, among so far identified 60 BRs-related compounds, the list of bioactive BRs includes only three: brassinolide (BL), 28-homobrassinolide (28-HomoBL), and 24-epibrassinolide (24-EpiBL) [60–62]. Extensive reports are available on the role of BRs in salinity-impact control in different test plants [62–67].

In several salinity-exposed test plants, 28-homoBL detoxified the NaCl-caused stress by elevating the activities of antioxidative enzymes including (SOD, CAT, GR, APX, and GPX) [68, 69]. Seed priming with BL can improve seed germination and seedling growth by significantly increasing POD, SOD, and CAT activity under salt stress [70]. The supply of polyhydroxylated spirostanic brassinosteroid analog (BB-16) can also enhance the activity of CAT, SOD, and GR and thereby mitigate the salinity impacts in plants [71]. 24-EpiBL-mediated improved tolerance of different test plants to varying salinity levels involved a 24-EpiBL-mediated decrease in oxidative stress via induction in the activity of ROS-metabolizing enzymatic antioxidants including APX, CAT, and POD [72–77]. Significant decreases in the cellular levels of electrolyte leakage, $O_2^{\bullet-}$ production, MDA, H_2O_2, and improved growth, carbonic anhydrase activity, photosynthetic efficiency in epiBL-supplied salinity-treated test plants were corroborated with enhanced activity of SOD, POD, GPX, CAT and APX enzymes and the improved contents of AsA and GSH [78, 79]. Interestingly, BRs have been extensively reported to regulate plant-salinity tolerance via interacting with a number of plant hormones including auxins [80, 81], ethylene [63, 65], ABA [82–85], and NO [64, 79, 82, 86, 87]. Moreover, BR signaling components can be directly regulated by salt stress signals at both transcriptional and post-translational levels [84, 85, 88–90].

3.4 Cytokinins

Cytokinins (CKs) are the derivatives of adenine (such as zeatin, kinetin, and N6-benzyladenine, BA) or of phenylurea (such as diphenylurea and thidiazuron) [45, 91]. Notably, the first naturally occurring CK was zeatin, which was identified and purified from immature maize (*Zea mays*) kernels [92]. Kinetin was the first CK discovered as an adenine (aminopurine) derivative [92]. CKs mainly regulate the major plant growth and developmental processes [93, 94]. However, literature also supports the immense roles (and underlying mechanisms) of CKs in plant abiotic stress tolerance [95–97]. Notably, the genetic engineering of CKs-metabolism was argued as one of the prospective ways to improve agricultural traits of crop plants [98]. CK signaling-mediated promotion in salt tolerance in *Z. mays* was argued to involve CK-mediated modulation of shoot Cl^- exclusion [99]. Soaking *Z. mays* seeds in zeatin-type cytokinin biostimulators (namely cis-zeatin-type CKs, c-Z-Ck; trans-zeatin, t-Z-Ck isomers) was reported to enhance antioxidant system and photosynthetic efficiency and thereby improve *Z. mays* salt tolerance [100]. Notable contradictory results yielded in several studies on the functional analyses of CK receptor mutants and the involvement of CK in ABA-mediated stress signaling in plants under osmotic/salinity stress warrant further molecular-genetic clarifications regarding the role of CKs in plant osmotic/salinity stress tolerance [101–103].

3.5 Ethylene

Ethylene is an important signaling molecule and a gaseous phytohormone. Its coordination with downstream signaling components has been reported to help plants in varyingly tolerating salinity stress [104–106]. Induction of ethylene generation in salinity-exposed plants is indicative of its significance as a downstream signal and modulation of gene expression [104]. Ethylene-homeostasis and ethylene signaling have been argued as an important factor required for plant-salinity tolerance [107, 108]. The maintenance of cellular ethylene (via endogenous production-induced accumulation and/or by exogenously supplied of ethylene precursor, 1-aminocyclopropane-1-carboxylic acid) has been reported to enhance Na^+ and K^+ homeostasis and induce downstream signaling for ROS-homeostasis; and eventually to improve plant salt tolerance [109–111]. Earlier, ethylene-mediated improvement in *Arabidopsis* salt tolerance mainly involved enhanced retention of K^+ in shoots and roots rather than decrease in tissue Na^+ content [112]. Ethylene (or its biochemical precursor, 1-aminocyclopropane-1-carboxylic acid) supplies improved plant tolerance to high salinity [110, 113–115]. Ethylene can trigger plant salt tolerance by modulating polyamine catabolism enzymes associated with H_2O_2 production [116]. S-nitrosylation of ACO homolog 4 (1-aminocyclopropane-1-carboxylate oxidase homolog 4; ACOh4) improved ethylene synthesis and improved salt tolerance in salinity-exposed tomato plants [117]. The roles of S-adenosylmethionine (SAM, involved in ethylene biosynthesis) and its derivatives in plant salt tolerance have also been recently discussed [118].

Molecular studies have unveiled the 'MdNAC047-ETHYLENE RESPONSE FACTOR (MdERF3)-ethylene-salt tolerance' regulatory pathway in apple [108]. Apple *MdERF4* was reported to negatively regulate salt tolerance by inhibiting *MdERF3* transcription [119]. *MdMYB46* enhanced salt (and osmotic) stress tolerance in apple by directly activating stress-responsive signals [120]. Additionally, plant responses to salt stress may also involve EIN3/EIL1-dependent genes and other ROS scavenger-coding genes [110]. The outcomes of crosstalk between *miR319* and ethylene contribute to plant-salinity tolerance. To this end, overexpression of *Osa-MIR319b* and targeting mimicry form of *miR319 (MIM319)* confirmed the role of *miR319*-mediated positive regulation of ethylene synthesis, and eventually improved salinity tolerance in switchgrass (*Panicum virgatum*) [121]. However, negative roles of ethylene have also been reported in salinity-exposed plants, where enhanced ethylene levels did not help plants in counteracting salinity stress impacts [122–124]. Thus, the reported few ambiguous roles of ethylene in plant-salinity stress responses require further explanations.

3.6 Gibberellins

Gibberellins (gibberellic acid, GA), a large family of tetracyclic di-terpenoid compounds, are classical plant hormones denoted largely by 'gibberellin numbers' (GA_n) in order of discovery, such as GA1, GA2, ..., GA_n. In general, GAs are involved in growth and development [125, 126]. However, the literature is full on the involvement of GA in plant tolerance to a number of abiotic stresses [96, 126, 127]. In salinity-exposed barley, exogenous GA3 increased the shoot and root length of germinated barley seeds; significantly reduced ion-leakage, osmolyte (proline) accumulation; and thereby rescued the expression of the *HvABI5*, *HvABA7*, and *HvKO1* by 3, 10, and 33 fold, respectively [128]. Exogenously applied GA enhanced growth and salinity stress tolerance in *Z. mays* by modulating the morpho-physiological, biochemical, and molecular attributes [129]. Moreover, GA3-supply improved pigment content, plant growth, and development, reduced Na^+ concentration in shoots and roots, increased the water absorption and metabolic activities in seeds, uplifted the seed dormancy, modulated cell division, and cell elongation. In this way, GA3-supply increased the growth of root, shoot, and number of leaves; increased photosynthetic activities and the dry matter production; maintained a fine-tuning among AsA-GSH cycle components; improved the plant height, yield, and yield-related traits, Ca^{2+} and K^+ concentrations, and transpiration rates; and decreased Na^+ concentrations in different test plants under salinization [129–133].

As reported in plants under most abiotic stresses, a higher accumulation of DELLA proteins was reported in salinity-exposed plants, which in turn was argued to restrain growth and enhance stress tolerance through reducing GA signaling activity [134, 135]. Seed priming with GA was reported to induce high salinity tolerance in *Pisum sativum*, where the applied GA modulated antioxidants, secondary metabolites, and upregulated antiporter genes [136]. Moreover, pre-treating/soaking of seeds with GAs was widely evidenced to improve increased α-amylase; salinity-caused nutritional disorders; decreased Na^+ content; enhanced ion uptake, photosynthesis, and redox homeostasis; improved coordination among CAT, APX, and SOD, as an adaptive mechanism to salt stress [137–140].

3.7 Jasmonic acid

Important critical signaling molecule jasmonic acid (JA) is among the most abundant members of the jasmonate class of plant hormones. Derived from linolenic acid (as cyclopentanone) and lipids, JA is known to regulate plant growth, development, and stress responses [45, 141]. Extensive reports are available on the role (and underlying mechanisms) of JA-mediated plant-salinity tolerance. Methyl jasmonic acid supply shifted the endogenous fatty acid levels and supported *O. sativa* growth in saline soil [142]. Transcriptomic analysis has revealed methyl jasmonate-mediated salt tolerance in alfalfa (*Medicago sativa*) as a result of antioxidant activity regulation and ion homeostasis [99]. JA-supply mediated mitigation of the inhibitory effect of salt stress in *T. aestivum* by increasing the endogenous levels of CK and IAA, reducing ABA contents, increasing α-tocopherol, phenolics, and flavonoids levels, and triggering SOD and APX activity [143]. In salinity-exposed *Anchusa italica*, methyl jasmonate improved test plant-salinity tolerance by enhancing contents of photosynthetic pigment, soluble sugars, K^+ and Ca^{2+}, declining Na^+ content, and eventually improving the major growth attributes [144]. Salinity stress-mediated induction in endogenous JA is also known in plants [145, 146]. Interaction outcomes of JA with ABA can also improve plant-salinity tolerance [141, 147]. JA-mediated saline stress tolerance in *O. sativa* involved autophagy and programmed cell death as critical pathways [148]. Jasmonate biosynthesis gene *OsOPR7* was involved in the mitigation of salinity-induced mitochondrial oxidative stress [149]. Conferment of a greater cell elongation under salt stress was achieved with mutations of JA-receptor CORONATINE

INSENSITIVE1 (COI1) and MYC2/3/4 along with the stabilized JASMONATE ZIM mutant jaz3-1 [150].

3.8 Nitric oxide

A highly versatile gaseous, free-radical, redox-signaling molecule, nitric oxide (NO), has been widely reported to perform diverse functions in plants [15, 151]. In a wide range of studies on salinity-exposed plants, exogenous supply of NO (or sodium nitroprusside, SNP; a NO donor) improved seed vigor, germination, and plant health and productivity through alleviating oxidative damage as a result of decreased levels of electrolyte leakage, MDA, and H_2O_2, improving antioxidant defense mechanism, decreasing methylglyoxal toxicity, and upregulating the glyoxalase system, adjusting the levels of osmolytes, and maintaining ionic balance [16, 152–154]. NO-supply can also reverse the glucose-mediated photosynthetic repression in plants under salinity exposure [154]. In NO (0.1 mM SNP)-mediated *T. aestivum* seed priming (for 20 h) helped improve germination rate, the weight of radical and coleoptile, and K^+ ion and Na^+ ion homeostasis [155, 156]. Exogenous application of NO (or its donor, SNP) can increase leaf area, plant dry mass, and the lengths of shoot and root in NaCl-stressed plants [157, 158]. Additionally, the maintenance of ion homeostasis (via enhanced K^+ uptake and reduced Na^+ uptake) and the modulation of the Na^+/H^+ antiporter enzyme were also reported in salinity-exposed and NO (or its donor, SNP)-supplemented test plants [159]. NO-mediated high salt tolerance in plants may also involve a NO-accrued increase in H^+-ATPase activity and eventual reduced leakage-mediated maintenance of high cytosolic K^+/Na^+ ratio [156, 160]. NO-mediated improved defense against salt-induced stress also involves NO's interaction with signaling molecules [15]. In earlier studies, NO (or its donor, SNP)-supply mediated improvements in the photosynthetic capacity involved NO-mediated protection of photosynthetic pigments; maintenance of normal shape of thylakoids and increase in chloroplast size; enhancement in the quenching of additional energy and quantum-yield of photosystem II; increase in ribulose-1,5-bisphosphate carboxylase/oxygenase (RuBisCO) activity; induction in the influx and efflux of Ca^{2+}; regulation of stomatal behavior and guard cells-ABA concentration; and efficient energy dissipation [158, 161–163].

The asA-GSH cycle is a central modulator of the plant stress responses and defense, ROS metabolism, and cellular redox balance [26, 164]. A plethora of reports supports the role of NO in the maintenance of the cellular redox balance via regulation of AsA-GSH cycle components (enzymatic and non-enzymatic) in salinity-impacted plants [123, 165, 166]. In different salinity-exposed plants, NO (or its donor, SNP)-supply resulted in significantly decreased cellular $O_2^{\bullet-}$ generation, H_2O_2, MDA content, and electrolyte leakage via maintaining a fine-tuning among antioxidant enzymes (including SOD, CAT, APX a H_2O_2-scavenging enzyme, MDHAR, DHAR, GR, GST, GPX, and CAT) and non-enzymatic antioxidants (including GSH and AsA) [15, 16, 151, 167].

3.9 Salicylic acid

Salicylic acid (SA), a phenolic plant hormone, is widely known for its involvement in plant growth and development and modulation of plant stress responses. Exogenous SA supply has strengthened salinity stress-tolerance mechanisms in extensive studies [168, 169]. Reports are also available on a high bio-stimulatory capacity of SA for salinity tolerance involving positive regulation of the AsA-GSH cycle, elevated accumulation of osmoprotectors, antioxidant enzyme activation, and increasing tolerance under ion toxicity and oxidative stress [34, 170–172]. In salinity-exposed *Vigna radiata*, the role of SA-induced accumulation of glycine betaine protected photosynthesis and growth against NaCl-accrued impacts in *V. radiata* as a result of the minimized accumulation of Na^+ and Cl^- ions and oxidative stress and maintained high GSH level and eventually reduced cellular redox environment [25]. In many instances, SA-mediated plant-salinity tolerance has SA-dose dependency [25, 40, 173, 174]. In salinity (100 mM NaCl)-exposed pepper (*Capsicum annuum*) plants, an exogenous supply of SA (0.5 mM) reduced leaf Na^+ content and oxidative stress-related traits [171]. SA-supply mediated regulation of ROS-metabolism and AsA-GSH cycle has also been reported in plants under salinity stress [175]. In earlier studies, supplied SA-assisted mitigation of salinity stress impacts in plants involved characteristic changes in the expression pattern of GST-gene family members such as *SlGSTT2, SlGSTT3,* and *SlGSTF4* [176]; enhanced transcript level of antioxidant genes; *GPX1, GPX2, DHAR, GR, GST1, GST2, MDHAR,* and *GS* [177], and GORK channel-mediated control of K^+ loss [178].

3.10 Strigolactones (SLs)

Synthesized in several plant species, strigolactones (SLs) are multifunctional β-carotene derivative molecules and are considered as an essential plant hormone in regulating plant functions. SLs have been considered an emerging growth regulator for developing resilience in plants [37, 41, 179]. In salinity-treated rapeseed (*Brassica napus*), the supplied SL increased plant growth, photosynthetic traits, and antioxidant enzyme activity [180]. Involvement of SL signal transduction and SL-biosynthetic mutants in *more axillary growth3 (max3)* and *max4* was reported in SLs-mediated positive regulation of plant salt tolerance [181, 182]. In many instances, SL-mediated plant salt tolerance involved ABA accumulation [40, 183, 184]. In some studies on salinity-exposed plants, exogenous SLs improved ROS metabolism and decreased lipid peroxidation and cellular damage [185, 186]. The role of arbuscular mycorrhizal fungi and ABA in SL-assisted improvement in plant-salinity tolerance has also been reported [183, 184].

4. Conclusions and prospects

The salinity of soils is significantly limiting crop production in most arid and semi-arid regions of the world. Owing to their sensitivity to salinity stress, most agricultural crop plants exhibit soil salinization-accrued inhibition in growth, metabolism, development, and productivity (yield). Thus, salinity-caused serious impacts on crop health and productivity are challenging food security. Though very small molecules are produced in small quantities, phytohormones have versatile roles in plants as the major regulator of various signaling cascades interrelated with plant development, stress resilience, and coping mechanisms. Apart from presenting a brief overview, this chapter attempted to enlighten the major roles (and the basic mechanisms involved) of selected 10 phytohormones (ABA, auxins, BRs, CKs, ethylene, GAs, JA, NO, SA, and SLs) in plant-salinity tolerance. An optimum concentration and timely exogenous application of these phytohormones would be a promising approach for countering the salinity stress impacts in crop plants. There are still challenges to understanding how ABA, auxins, BRs, CKs, ethylene, GAs, JA, NO, SA, and SLs and associated close molecules can function at the molecular level and how the intimate mechanisms of interaction among these phytohormones and also with other emerging signaling molecules work. Additionally, a few contradictory results related to the involvement of CK in ABA-mediated stress signaling and the reported few ambiguous roles of ethylene in plant-salinity stress responses under osmotic/salinity stress warrant further molecular-genetic clarifications.

Author details

Naser A. Anjum[1*], Asim Masood[1], Faisal Rasheed[1], Palaniswamy Thangavel[2] and Nafees A. Khan[1]

1 Department of Botany, Aligarh Muslim University, Aligarh, India

2 Department of Environmental Science, Periyar University, Salem, India

*Address all correspondence to: naanjum1@myamu.ac.in

References

[1] Tigchelaar M, Battisti DS, Naylor RL, Ray DK. Future warming increases probability of globally synchronized maize production shocks. Proceedings of the National Academy of Sciences, USA. 2018;**115**:644-6649

[2] Vaughan MM, Block A, Christensen SA, Allen LH, Schmelz EA. The effects of climate change associated abiotic stresses on maize phytochemical defenses. Phytochemistry Reviews. 2018;**17**:37-49

[3] Zörb C, Geilfus CM, Dietz KJ. Salinity and crop yield. Plant Biology. 2019;**21**:31-38

[4] Intergovernmental Panel on Climate Change (IPCC). Climate Change 2022. Impacts, adaptation and vulnerability. In: Pörtner HO, Roberts DC, Tignor M, et al., editors. Contribution of Working Group II to the Sixth Assessment Report of the Intergovernmental Panel on Climate Change. UK and New York: Cambridge University Press; 2022. DOI: 10.1017/9781009325844

[5] Zeifman L, Hertog S, Kantorova V, Wilmoth J. A World of 8 Billion. Policy Brief No. 140, Population Division, UN Department of Economic and Social Affairs. 2022. pp. 1-4. Available from: https://www.un.org/development/desa/dpad/wp-content/uploads/sites/45/publication/PB_140.pdf

[6] Cassidy ES, West PC, Gerber JS, Foley JA. Redefining agricultural yields: From tonnes to people nourished per hectare. Environmental Research Letters. 2013;**8**(3):034015

[7] Hemathilake DM, Gunathilake DM. Agricultural productivity and food supply to meet increased demands. In: Future Foods. Cambridge, Massachusetts, United States: Academic Press; 2022. pp. 539-553

[8] Satir O, Berberoglu S. Crop yield prediction under soil salinity using satellite derived vegetation indices. Field Crops Research. 2016;**192**:134-143

[9] Wicke B, Smeets E, Dornburg V, Vashev B, Gaiser T, Turkenburg W, et al. The global technical and economic potential of bioenergy from salt-affected soils. Energy and Environmental Science. 2011;**4**:26692681

[10] Ayub MA, Ahmad HR, Ali M, Rizwan M, Ali S, Rehman MZ, et al. Salinity and its tolerance strategies in plants. In: Plant Life under Changing Environment. Cambridge, Massachusetts, United States: Academic Press; 2020. pp. 47-76

[11] Rasheed F, Anjum NA, Masood A, Sofo A, Khan NA. The key roles of salicylic acid and sulfur in plant salinity stress tolerance. Journal of Plant Growth Regulation. 2022;**41**:1891-1904

[12] Rasheed F, Sehar Z, Fatma M, Iqbal N, Masood A, Anjum NA, et al. Involvement of ethylene in reversal of salt stress by salicylic acid in the presence of sulfur in mustard (*Brassica juncea* L.). Journal of Plant Growth Regulation. 2022;**41**:3449-3466

[13] Rengasamy P. Soil processes affecting crop production in salt-affected soils. Functional Plant Biology. 2010;**37**:613-620

[14] Ludwiczak A, Osiak M, Cárdenas-Pérez S, Lubińska-Mielińska S, Piernik A. Osmotic stress or ionic

composition: Which affects the early growth of crop species more? Agronomy. 2021;**11**(3):435

[15] Hasanuzzaman M, Oku H, Nahar K, Bhuyan MB, Mahmud JA, Baluska F, et al. Nitric oxide-induced salt stress tolerance in plants: ROS metabolism, signaling, and molecular interactions. Plant Biotechnology Reports. 2018;**12**:77-92

[16] Hasanuzzaman M, Hossain MA, Fujita M. Nitric oxide modulates antioxidant defense and methylglyoxal detoxification system and reduces salinity induced damage in wheat seedling. Plant Biotechnology Reports. 2011;**5**:353-365

[17] Food and Agricultural Organization (FAO). Global Soil Resources. 2015. Available from: https://hal.archives-ouvertes.fr/hal-01241064/

[18] Hossain MS. Present scenario of global salt affected soils, its management and importance of salinity research. International Research Journal of Biological Sciences. 2019;**1**:1-3

[19] Qadir M, Quillerou E, Nangia V, Murtaza G, Singh M, Thomas RJ, et al. Economics of salt-induced land degradation and restoration. Natural Resources Forum. 2014;**38**:282-295

[20] Shabala S, Wu H, Bose J. Salt stress sensing and early signalling events in plant roots: Current knowledge and hypothesis. Plant Science. 2015;**241**:109-119

[21] Singh AK, Gupta KJ, Singla-Pareek SL, Foyer CH, Pareek A. Raising crops for dry and saline lands: Challenges and the way forward. Physiologia Plantarum. 2022;**174**:1-4

[22] Wang CF, Han GL, Yang ZR, Li YX, Wang BS. Plant salinity sensors: Current understanding and future directions. Frontiers in Plant Science. 2022;**13**:859224

[23] Awlia M, Alshareef N, Saber N, Korte A, Oakey H, Panzarová K, et al. Genetic mapping of the early responses to salt stress in *Arabidopsis thaliana*. The Plant Journal. 2021;**107**(2):544-563

[24] Yang Y, Guo Y. Unraveling salt stress signaling in plants. Journal of Integrative Plant Biology. 2018;**60**:796-804

[25] Syeed S, Sehar Z, Masood A, Anjum NA, Khan NA. Control of elevated ion accumulation, oxidative stress, and lipid peroxidation with salicylic acid-induced accumulation of glycine betaine in salinity-exposed *Vigna radiata* L. Applied Biochemistry and Biotechnology. 2021;**193**(10):3301-3320

[26] Anjum NA, Umar S, Ahmad A. Oxidative Stress in Plants: Causes, Consequences and Tolerance. New Delhi: IK International Publishing House; 2012

[27] Anjum NA, Sofo A, Scopa A, Roychoudhury A, Gill SS, Iqbal M, et al. Lipids and proteins - major targets of oxidative modifications in abiotic stressed plants. Environmental Science and Pollution Research. 2015;**22**:4099-4121

[28] Mittler R. Oxidative stress, antioxidants and stress tolerance. Trends in Plant Science. 2002;**7**:405-410

[29] Gill SS, Tuteja N. Reactive oxygen species and antioxidant machinery in abiotic stress tolerance in crop plants. Plant Physiology and Biochemistry. 2010;**48**:909-930

[30] Anjum NA, Gill SS, Gill R, Hasanuzzaman M, Duarte AC, Pereira E, et al. Metal/metalloid stress tolerance in plants: Role of ascorbate, its redox

couple, and associated enzymes. Protoplasma. 2014;**251**:1265-1283

[31] Anjum NA, Sharma P, Gill SS, Hasanuzzaman M, Khan EA, Kachhap K, et al. Catalase and ascorbate peroxidase-representative H_2O_2-detoxifying heme enzymes in plants. Environmental Science and Pollution Research. 2016;**23**:19002-19029

[32] Anjum NA, Gill SS, Corpas FJ, Ortega-Villasante C, Hernandez LE, Tuteja N, et al. Editorial: Recent insights into the double role of hydrogen peroxide in plants. Frontiers in Plant Science. 2022;**13**:843274

[33] Gill SS, Anjum NA, Gill R, Yadav S, Hasanuzzaman M, Fujita M, et al. Superoxide dismutase – Mentor of abiotic stress tolerance in crop plants. Environmental Science and Pollution Research. 2015;**22**:10375-10394

[34] Hasanuzzaman M, Bhuyan MB, Anee TI, Parvin K, Nahar K, Mahmud JA, et al. Regulation of ascorbate-glutathione pathway in mitigating oxidative damage in plants under abiotic stress. Antioxidants. 2019;**8**(9):384

[35] Khan NA, Nazar R, Iqbal N, Anjum NA. Phytohormones and Abiotic Stress Tolerance in Plants. Berlin, Heidelberg, Germany: Springer-Verlag; 2012

[36] Ryu H, Cho YG. Plant hormones in salt stress tolerance. Journal of Plant Biology. 2015;**58**:147-155

[37] El-Esawi MA. Introductory chapter: Hormonal regulation in plant development and stress tolerance. In: Phytohormones-Signaling Mechanisms and Crosstalk in Plant Development and Stress Responses. London, UK: IntechOpen; 2017. pp. 1-9

[38] Sharma A, Shahzad B, Kumar V, Kohli SK, Sidhu GP, Bali AS, et al. Phytohormones regulate accumulation of osmolytes under abiotic stress. Biomolecules. 2019;**9**(7):285

[39] Verma V, Ravindran P, Kumar PP. Plant hormone-mediated regulation of stress responses. BMC Plant Biology. 2016;**16**:86

[40] Yu Z, Duan X, Luo L, Dai S, Ding Z, Xia G. How plant hormones mediate salt stress responses. Trends in Plant Science. 2020;**25**(11):1117-1130

[41] Wani SH, Kumar V, Shriram V, Sah SK. Phytohormones and their metabolic engineering for abiotic stress tolerance in crop plants. The Crop Journal. 2016;**4**:162-176

[42] Sreenivasulu N, Harshavardhan VT, Govind G, Seiler C, Kohli A. Contrapuntal role of ABA: Does it mediate stress tolerance or plant growth retardation under long-term drought stress? Gene. 2012;**506**:265-273

[43] Sah SK, Reddy KR, Li J. Abscisic acid and abiotic stress tolerance in crop plants. Frontiers in Plant Science. 2016;7:571

[44] Cutler SR, Rodriguez PL, Finkelstein RR, Abrams SR. Abscisic acid: Emergence of a core signaling network. Annual Review of Plant Biology. 2010;**61**:651-679

[45] Davies PJ. Plant Hormones: Biosynthesis, Signal Transduction, Action! 3rd ed. Dordrecht: Springer; 2010

[46] Ruggiero A, Landi S, Punzo P, Possenti M, Van Oosten MJ, Costa A, et al. Salinity and ABA seed responses in pepper: Expression and interaction

of ABA core signaling components. Frontiers in Plant Science. 2019;**10**:304

[47] Majid A, Rather BA, Masood A, Sehar Z, Anjum NA, Khan NA. Abscisic acid in coordination with nitrogen alleviates salinity-inhibited photosynthetic potential in mustard by improving proline accumulation and antioxidant activity. Stress. 2021;**1**(3):162-180

[48] Chen G, Zheng D, Feng N, Zhou H, Mu D, Zhao L, et al. Physiological mechanisms of ABA-induced salinity tolerance in leaves and roots of rice. Scientific Reports. 2022;**12**(1):8228

[49] Parveen A, Ahmar S, Kamran M, Malik Z, Ali A, Riaz M, et al. Abscisic acid signaling reduced transpiration flow, regulated Na^+ ion homeostasis and antioxidant enzyme activities to induce salinity tolerance in wheat (*Triticum aestivum* L.) seedlings. Environmental Technology and Innovation. 2021;**24**:101808

[50] Sripinyowanich S, Klomsakul P, Boonburapong B, Bangyeekhun T, Asami T, et al. Exogenous ABA induces salt tolerance in Indica rice (*Oryza sativa* L.): The role of *OsP5CS1* and *OsP5CR* gene expression during salt stress. Environmental and Experimental Botany. 2013;**86**:94-105

[51] Carrillo-Carrasco VP, Hernandez-Garcia J, Mutte SK, Weijers D. The birth of a giant: Evolutionary insights into the origin of auxin responses in plants. The EMBO Journal. 2023;**42**(6):e113018

[52] Jing H, Wilkinson EG, Sageman-Furnas K, Strader LC. Auxin and abiotic stress responses. Journal of Experimental Botany. 2023;**17**:erad325

[53] Zhao Y. Auxin biosynthesis and its role in plant development. Annual Review of Plant Biology. 2010;**61**:49-64

[54] Weijers D, Nemhauser J, Yang Z. Auxin, small molecule, big impact. Journal of Experimental Botany. 2018;**5**:133-136

[55] Iglesias MJ, Terrile MC, Windels D, Lombardo MC, Bartoli CG, Vazquez F, et al. MiR393 regulation of auxin signaling and redox-related components during acclimation to salinity in *Arabidopsis*. PLoS One. 2014;**9**:e107678

[56] Liu W, Li RJ, Han TT, Cai W, Fu ZW, Lu YT. Salt stress reduces root meristem size by nitric oxide-mediated modulation of auxin accumulation and signaling in *Arabidopsis*. Plant Physiology. 2015;**168**:343-356

[57] Petersson SV, Johansson AI, Kowalczyk M, Makoveychuk A, Wang JY, et al. An auxin gradient and maximum in the *Arabidopsis* root apex shown by high-resolution cell-specific analysis of IAA distribution and synthesis. The Plant Cell. 2009;**21**:1659-1668

[58] Wang Y, Li K, Li X. Auxin redistribution modulates plastic development of root system architecture under salt stress in *Arabidopsis thaliana*. Journal of Plant Physiology. 2009;**166**:1637-1645

[59] Akbari G, Sanavy SA, Yousefzadeh S. Effect of auxin and salt stress (NaCl) on seed germination of wheat cultivars (*Triticum aestivum* L.). Pakistan Journal of Biological Sciences. 2007;**10**:2557-2561

[60] Rao SSR, Vardhini BV, Sujatha E, Anuradha S. Brassinosteroids – New class of phytohormones. Current Science. 2002;**82**:1239-1245

[61] Vriet C, Russinova E, Reuzeau C. Boosting crop yields with plant steroids. The Plant Cell. 2012;**24**:842-857

[62] Vardhini BV, Anjum NA. Brassinosteroids make plant life easier under abiotic stresses mainly by modulating major components of antioxidant defense system. Frontiers in Environmental Science. 2015;**2**:67

[63] Zhu T, Deng X, Zhou X, Zhu L, Zou L, Li P, et al. Ethylene and hydrogen peroxide are involved in brassinosteroid-induced salt tolerance in tomato. Scientific Reports. 2016;**6**:35392

[64] Zhu T, Deng XG, Tan WR, Zhou X, Luo SS, Han XY, et al. Nitric oxide is involved in brassinosteroid-induced alternative respiratory pathway in *Nicotiana benthamiana* seedlings' response to salt stress. Physiologia Plantarum. 2016;**156**:150-163

[65] Wei LJ, Deng XG, Zhu T, Zheng T, Li PX, Wu JQ, et al. Ethylene is involved in brassinosteroids induced alternative respiratory pathway in cucumber (*Cucumis sativus* L.) seedlings response to abiotic stress. Frontiers in Plant Science. 2015;**6**:982

[66] Su Q, Zheng X, Tian Y, Wang C. Exogenous brassinolide alleviates salt stress in *Malus hupehensis* Rehd. by regulating the transcription of NHX-type $Na^+(K^+)/H^+$ antiporters. Frontiers in Plant Science. 2020;**11**:38

[67] Yan J, Guan L, Sun Y, Zhu Y, Liu L, Lu R, et al. Calcium and ZmCCaMK are involved in brassinosteroid-induced antioxidant defense in maize leaves. Plant and Cell Physiology. 2015;**56**:883-896

[68] Arora N, Bhardwaj R, Sharma P, Arora HK. Effects of 28-homobrassinolide on growth, lipid peroxidation and antioxidative enzyme activities in seedlings of *Zea mays* L. under salinity stress. Acta. Physiologia Plantarum. 2008;**30**:833-839

[69] Hayat S, Hasan SA, Yusuf M, Hayat Q, Ahmad A. Effect of 28-homobrassinolide on photosynthesis, fluorescence and antioxidant system in the presence or absence of salinity and temperature in *Vigna radiata*. Environmental and Experimental Botany. 2010;**69**:105-112

[70] Zhang S, Hu J, Zhang Y, Xie XJ, Knapp A. Seed priming with brassinolide improves lucerne (*Medicago sativa* L.) seed germination and seedling growth in relation to physiological changes under salinity stress. Australian Journal of Agricultural Research. 2007;**58**:811-815

[71] Nunez M, Mazzafera P, Mazorra LM, Siqueira WJ, Zullo MAT. Influence of a brassinosteroid analog on antioxidant enzymes in rice grown in culture medium with NaCl. Biologia Plantarum. 2003;**47**:67-70

[72] Özdemir F, Bor M, Demiral T, Türkan I. Effects of 24-epibrassinolide on seed germination, seedling growth, lipid peroxidation, proline content and antioxidative system of rice (*Oryza sativa* L.) under salinity stress. Plant Growth Regulation. 2004;**42**:203-211

[73] Shahbaz M, Ashraf M. Influence of exogenous application of brassinosteroids on growth and mineral nutrients of wheat (*Triticum aestivum* L.) under saline conditions. Pakistan Journal of Botany. 2007;**39**:513-522

[74] Ali B, Hasan SA, Hayat S, Hayat Q, Yadav S, Fariduddin Q, et al. A role for brassinosteroids in the amelioration of aluminium stress through antioxidant system in mung bean (*Vigna radiata* L. Wilczek). Environmental and Experimental Botany. 2008;**62**:153-159

[75] Shahid MA, Pervez MA, Balal RM, Mattson NS, Rashid A, Ahmad R, et al. Brassinosteroid (24-epibrassinolide) enhances growth and alleviates the deleterious effects induced by salt stress in pea (*Pisum sativum* L.). Austrian Journal of Crop Science. 2011;**5**:500-510

[76] Sharma I, Ching E, Saini S, Bhardwaj R, Pati PK. Exogenous application of brassinosteroid offers tolerance to salinity by altering stress responses in rice variety Pusa Basmati-1. Plant Physiology and Biochemistry. 2013;**69**:17-26

[77] Abbas S, Latif HH, Elsherbiny EA. Effect of 24-epibrassinolide on the physiological and genetic changes on two varieties of pepper under salt stress conditions. Pakistan Journal of Botany. 2013;**45**:1273-1284

[78] Ding HD, Zhu XH, Zhu ZW, Yang SJ, Zha DS, et al. Amelioration of salt-induced oxidative stress in eggplant by application of 24-epibrassinolide. Biologia Plantarum. 2012;**56**:767-770

[79] Fariduddin Q, Mir BA, Yusuf M, Ahmad A. Comparative roles of brassinosteroids and polyamines in salt stress tolerance. Acta Physiologia Plantarum. 2013;**35**:2037-2053

[80] Mouchel CF, Osmont KS, Hardtke CS. BRX mediates feedback between brassinosteroid levels and auxin signalling in root growth. Nature. 2006;**443**(7110):458-461

[81] Djemal R, Hanin M, Ebel C. Control of plant responses to salt stress: Significance of auxin and brassinosteroids. In: Anjum NA, Masood A, Thangavel P, Khan NA, editors. Making Plant Life Easier and Productive Under Salinity – Updates and Prospects. London, UK: IntechOpen; 2023. DOI: 10.5772/intechopen.111449

[82] Zhang S, Cai Z, Wang X. The primary signaling outputs of brassinosteroids are regulated by abscisic acid signaling. Proceedings of the National Academy of Sciences, USA. 2009;**106**:4543-4548

[83] Zeng H, Tang Q, Hua X. Arabidopsis brassinosteroid mutants det2-1 and bin2-1 display altered salt tolerance. Journal of Plant Growth Regulation. 2010;**29**:44-52

[84] Li ZY, Xu ZS, He GY, Yang GX, Chen M, Li LC, et al. A mutation in *Arabidopsis* BSK5 encoding a brassinosteroid-signaling kinase protein affects responses to salinity and abscisic acid. Biochemical and Biophysical Research Communications. 2012;**426**:522-527

[85] Sun F, Yu H, Qu J, Cao Y, Ding L, Feng W, et al. Maize ZmBES1/BZR1-5 decreases ABA sensitivity and confers tolerance to osmotic stress in transgenic *Arabidopsis*. International Journal of Molecular Sciences. 2020;**21**:996

[86] Wang X, Chen X, Wang Q, Chen M, Liu X, Gao D, et al. MdBZR1 and MdBZR1-2like transcription factors improves salt tolerance by regulating gibberellin biosynthesis in apple. Frontiers in Plant Science. 2019;**10**:1473

[87] Divi UK, Rahman T, Krishna P. Brassinosteroid-mediated stress tolerance in *Arabidopsis* shows interactions with abscisic acid, ethylene and salicylic acid pathways. BMC Plant Biology. 2010;**10**:1-4

[88] Geng Y, Wu R, Wee CW, Xie F, Wei X, Chan PM, et al. A spatio-temporal understanding of growth regulation during the salt stress response in *Arabidopsis*. The Plant Cell. 2013;**25**:2132-2154

[89] Duan F, Ding J, Lee D, Lu X, Feng Y, Song W. Overexpression of *SoCYP85A1*, a spinach cytochrome p450 gene in transgenic tobacco enhances root development and drought stress tolerance. Frontiers in Plant Science. 2017;**8**:1909

[90] Srivastava M, Srivastava AK, Orosa-Puente B, Campanaro A, Zhang C, Sadanandom A. SUMO conjugation to *BZR1* enables brassinosteroid signaling to integrate environmental cues to shape plant growth. Current Biology. 2020;**30**:1410-1423

[91] George E, Hall M, Klerk GJ. Plant growth regulators II: Cytokinins, their analogues and antagonists. In: George E, Hall M, Klerk GJ, editors. Plant Propagation by Tissue Culture. Dordrecht, The Netherlands: Springer; 2008. pp. 205-226

[92] Miller CO. A kinetin-like compound in maize. Proceeding of National Academy of Sciences, USA. 1961;**47**:170-174

[93] Mok DW, Mok MC. Cytokinin metabolism and action. Annual Review of Plant Physiology and Plant Molecular Biology. 2001;**52**:89-118

[94] Kieber JJ, Schaller GE. Cytokinin signaling in plant development. Development. 2018;**145**(4):dev149344

[95] O'Brien JA, Benková E. Cytokinin cross-talking during biotic and abiotic stress responses. Frontiers in Plant Science. 2013;**4**:451

[96] Asgher M, Khan MIR, Anjum NA, et al. Minimising toxicity of cadmium in plants - role of plant growth regulators. Protoplasma. 2015;**252**:399-413

[97] Cortleven A, Leuendorf JE, Frank M, Pezzetta D, Bolt S, Schmülling T. Cytokinin action in response to abiotic and biotic stresses in plants. Plant, Cell & Environment. 2019;**42**(3):998-1018

[98] Zalabák D, Pospíšilová H, Šmehilová M, Mrízová K, Frébort I, Galuszka P. Genetic engineering of cytokinin metabolism: Prospective way to improve agricultural traits of crop plants. Biotechnology Advances. 2013;**31**:97-117

[99] Yin P, Liang X, Zhao H, Xu Z, Chen L, Yang X, et al. Cytokinin signaling promotes salt tolerance by modulating shoot chloride exclusion in maize. Molecular Plant. 2023;**16**(6):1031-1047

[100] Azzam CR, Zaki SN, Bamagoos AA, Rady MM, Alharby HF. Soaking maize seeds in zeatin-type cytokinin biostimulators improves salt tolerance by enhancing the antioxidant system and photosynthetic efficiency. Plants. 2022;**11**(8):1004

[101] Merchan F, de Lorenzo L, Rizzo SG, Niebel A, Manyani H, Frugier F, et al. Identification of regulatory pathways involved in the reacquisition of root growth after salt stress in *Medicago truncatula*. Plant Journal. 2007;**51**:1-17

[102] Tran LS, Shinozaki K, Yamaguchi-Shinozaki K. Role of cytokinin responsive two-component system in ABA and osmotic stress signalings. Plant Signaling and Behavior. 2010;**5**:148-150

[103] Nishiyama R, Watanabe Y, Fujita Y, Le DT, et al. Analysis of cytokinin mutants and regulation of cytokinin metabolic genes reveals important regulatory roles of cytokinins in drought, salt and abscisic acid responses, and abscisic acid biosynthesis. The Plant Cell. 2011;**23**:2169-2183

[104] Wang KLC, Li H, Ecker JR. Ethylene biosynthesis and signaling networks. Plant Cell (Suppl). 2002;**14**:S131-S151

[105] Yang C, Lu X, Ma B, Chen SY, Zhang JS. Ethylene signaling in rice and *Arabidopsis*: Conserved and diverged aspects. Molecular Plant. 2015;**8**:495-505

[106] Zhang M, Smith JA, Harberd NP, Jiang C. The regulatory roles of ethylene and reactive oxygen species (ROS) in plant salt stress responses. Plant Molecular Biology. 2016;**91**:651-659

[107] Cao YR, Chen SY, Zhang JS. Ethylene signaling regulates salt stress response: An overview. Plant Signaling and Behavior. 2008;**3**(10):761-763

[108] An JP, Yao JF, Xu RR, You CX, Wang XF, Hao YJ. An apple NAC transcription factor enhances salt stress tolerance by modulating the ethylene response. Physiologia Plantarum. 2018;**164**(3):279-289

[109] Jiang C, Belfield EJ, Cao Y, Smith JA, Harberd NP. An Arabidopsis soil salinity-tolerance mutation confers ethylene-mediated enhancement of sodium/potassium homeostasis. The Plant Cell. 2013;**25**:3535-3552

[110] Peng J, Li Z, Wen X, Li W, Shi H, Yang L, et al. Salt-induced stabilization of EIN3/EIL1 confers salinity tolerance by deterring ROS accumulation in *Arabidopsis*. PLoS Genetics. 2014;**10**(10):e1004664

[111] Yang C, Ma B, He SJ, Xiong Q, Duan KX, Yin CC, et al. MAOHUZI6/ETHYLENE INSENSITIVE3-LIKE1 and ETHYLENE INSENSITIVE3-LIKE2 regulate ethylene response of roots and coleoptiles and negatively affect salt tolerance in rice. Plant Physiology. 2015;**169**:148-165

[112] Yang L, Zu YG, Tang ZH. Ethylene improves *Arabidopsis* salt tolerance mainly via retaining K^+ in shoots and roots rather than decreasing tissue Na^+ content. Environmental and Experimental Botany. 2013;**86**:60-69

[113] Khan MIR, Asgher M, Khan NA. Alleviation of salt-induced photosynthesis and growth inhibition by salicylic acid involves glycinebetaine and ethylene in mungbean (*Vigna radiata* L.). Plant Physiology and Biochemistry. 2014;**80**:67-74

[114] Silva PO, Medina EF, Barros RS, Ribeiro DM. Germination of salt-stressed seeds as related to the ethylene biosynthesis ability in three *Stylosanthes* species. Journal of Plant Physiology. 2014;**171**:14-22

[115] Cao WH, Liu J, He XJ, Mu RL, Zhou HL, Chen SY, et al. Modulation of ethylene responses affects plant salt-stress responses. Plant Physiology. 2017;**143**:707-719

[116] Freitas VS, de Souza MR, Costa JH, de Oliveira DF, de Oliveira PS, et al. Ethylene triggers salt tolerance in maize genotypes by modulating polyamine catabolism enzymes associated with H_2O_2 production. Environmental and Experimental Botany. 2018;**145**:75-86

[117] Liu M, Wei JW, Liu W, Gong B. S-nitrosylation of ACO homolog 4 improves ethylene synthesis and salt tolerance in tomato. New Phytologist. 2023;**239**:159-173

[118] Yang L, Wang X, Zhao F, Zhang X, Li W, Huang J, et al. Roles of S-adenosylmethionine and its derivatives in salt tolerance of cotton. International Journal of Molecular Sciences. 2023;**24**(11):9517

[119] An JP, Zhang XW, Xu RR, You CX, Wang XF, Hao YJ. Apple MdERF4 negatively regulates salt tolerance by inhibiting MdERF3 transcription. Plant Science. 2018;**276**:181-188

[120] Chen K, Song M, Guo Y, Liu L, Xue H, Dai H, et al. MdMYB46 could enhance salt and osmotic stress tolerance in apple by directly activating stress-responsive signals. Plant Biotechnology Journal. 2019;**17**:2341-2355

[121] Liu Y, Li D, Yan J, Wang K, Luo H, Zhang W. MiR319 mediated salt tolerance by ethylene. Plant Biotechnology Journal. 2019;**17**(12):2370-2383

[122] Ma B, Chen SY, Zhang JS. Ethylene signaling in rice. Chinese Science Bulletin. 2010;**55**:2204-2210

[123] Chen D, Ma X, Li C, Zhang W, Xia G, Wang M. A wheat aminocyclopropane-1-carboxylate oxidase gene, *TaACO1*, negatively regulates salinity stress in *Arabidopsis thaliana*. Plant Cell Reports. 2014;**33**:1815-1827

[124] Tao JJ, Chen HW, Ma B, Zhang WK, Chen SY, Zhang JS. The role of ethylene in plants under salinity stress. Frontiers in Plant Science. 2015;**6**:1059

[125] Matsuoka M. Gibberellins signaling: How do plant cells respond to GA signals? Journal of Plant Growth Regulation. 2003;**22**:123-125

[126] Gupta R, Chakrabarty SK. Gibberellic acid in plant – Still a mystery unresolved. Plant Signaling and Behavior. 2013;**8**:e25504

[127] Zawaski C, Busov VB. Roles of gibberellin catabolism and signaling in growth and physiological response to drought and short-day photoperiods in *Populus* trees. PLoS One. 2014;**9**:e86217

[128] Uçarli C. Physiological and molecular effects of exogenous gibberellin (GA3) treatment on germination of barley seeds under salt stress. Adıyaman University Journal of Science. 2021;**11**(2):227-243

[129] Shahzad K, Hussain S, Arfan M, Hussain S, Waraich EA, Zamir S, et al. Exogenously applied gibberellic acid enhances growth and salinity stress tolerance of maize through modulating the morpho-physiological, biochemical and molecular attributes. Biomolecules. 2021;**11**(7):1005

[130] Tuna AL, Kaya C, Dikilitas M, Higgs D. The combined effects of gibberellic acid and salinity on some antioxidant enzyme activities, plant growth parameters and nutritional status in maize plants. Environmental and Experimental Botany. 2008;**62**:1-9

[131] Ghodrat V, Rousta MJ. Effect of priming with gibberellic acid (GA3) on germination and growth of corn (*Zea mays* L.) under saline conditions. International Journal of Agricultural and Crop Science. 2012;**4**:882-885

[132] Iqbal M, Ashraf M. Gibberellic acid mediated induction of salt tolerance in wheat plants: Growth, ionic partitioning, photosynthesis, yield and hormonal homeostasis. Environmental and Experimental Botany. 2013;**86**:76-85

[133] Iftikhar A, Ali S, Yasmeen T, Arif MS, Zubair M, Rizwan M, et al. Effect of gibberellic acid on growth, photosynthesis and antioxidant defense system of wheat under zinc oxide nanoparticle stress. Environmental Pollution. 2019;**254**:113109

[134] Achard P, Renou J, Berthome R, Harberd N, Genschik P. Plant DELLAs restrain growth and promote survival of adversity by reducing the levels of reactive oxygen species. Current Biology. 2008;**18**:656-660

[135] Colebrook EH, Thomas SG, Phillips AL, Hedden P. The role of gibberellin signalling in plant responses to abiotic stress. Journal of Experimental Biology. 2014;**217**(1):67-75

[136] Ahmad F, Kamal A, Singh A, Ashfaque F, Alamri S, Siddiqui MH, et al. Seed priming with gibberellic acid induces high salinity tolerance in *Pisum sativum* through antioxidants, secondary metabolites and up-regulation of antiporter genes. Plant Biology. 2021;**23**:113-121

[137] Li Z, Xu J, Gao Y, Wang C, Guo G, Luo Y, et al. The synergistic priming effect of exogenous salicylic acid and H_2O_2 on chilling tolerance enhancement during maize (*Zea mays* L.) seed germination. Frontiers in Plant Science. 2017;**8**:1153

[138] Iqbal H, Yaning C, Waqas M, Rehman HM, Sharee FM, Iqbal S. Hydrogen peroxide application improves quinoa performance by affecting physiological and biochemical mechanisms under water-deficit conditions. Journal of Agronomy and Crop Sciences. 2018;**204**:541-553

[139] Al-Harthi MM, Bafeel SO, El-Zohri M. Gibberellic acid and jasmonic acid improve salt tolerance in summer squash by modulating some physiological parameters symptomatic for oxidative stress and mineral nutrition. Plants. 2021;**10**:2768

[140] Rady MM, Talaat NB, Abdelhamid MT, Shawky BT, Desoky EM. Maize (*Zea mays* L.) grains extract mitigates the deleterious effects of salt stress on common bean (*Phaseolus vulgaris* L.) growth and physiology. Journal of Horticultural Science and Biotechnology. 2021;**94**:777-789

[141] Kazan K, Manners JM. JAZ repressors and the orchestration of phytohormone crosstalk. Trends in Plant Science. 2012;**17**(1):22-31

[142] Zainordin ZF, San CT, Ahmad A. Foliar sprays of methyl jasmonic acid shift the endogenous fatty acid levels to support rice plant growth in saline soil. Plant Stress. 2023;**13**:100221

[143] Sheteiwy MS, Ulhassan Z, Qi W, Lu H, AbdElgawad H, Minkina T, et al. Association of jasmonic acid priming with multiple defense mechanisms in wheat plants under high salt stress. Frontiers in Plant Science. 2022;**13**:886862

[144] Taheri Z, Vatankhah E, Jafarian V. Methyl jasmonate improves physiological and biochemical responses of *Anchusa italica* under salinity stress. South African Journal of Botany. 2020;**130**:375-382

[145] Kramell R, Miersch O, Atzorn R, Parthier B, Wasternack C. Octadecanoid-derived alteration of gene expression and the 'oxylipin signature' in stressed barley leaves. Implications for different signaling pathways. Plant Physiology. 2000;**123**:177-187

[146] Pedranzani H, Racagni G, Alemano S, Miersch O, Ramirez I, Pena-Cortes H, et al. Salt tolerant tomato plants show increased levels of jasmonic acid. Plant Growth Regulation. 2003;**41**:149-158

[147] Shinozaki K, Yamaguchi-Shinozaki K. Gene networks involved in

drought stress response and tolerance. Journal of Experimental Botany. 2007;**58**(2):221-227

[148] Khan MS, Hemalatha S. Autophagy and programmed cell death are critical pathways in Jasmonic acid mediated saline stress tolerance in *Oryza sativa*. Applied Biochemistry and Biotechnology. 2022;**194**(11):5353-5366

[149] Asfaw KG, Liu Q, Eghbalian R, Purper S, Akaberi S, Dhakarey R, et al. The jasmonate biosynthesis *Gene OsOPR7* can mitigate salinity induced mitochondrial oxidative stress. Plant Science. 2022;**316**:111156

[150] Valenzuela CE, Acevedo-Acevedo O, Miranda GS, Vergara-Barros P, Holuigue L, Figueroa CR, et al. Salt stress response triggers activation of the jasmonate signaling pathway leading to inhibition of cell elongation in Arabidopsis primary root. Journal of Experimental Botany. 2016;**67**:4209-4220

[151] Rather BA, Masood A, Sehar Z, Majid A, Anjum NA, Khan NA. Mechanisms and role of nitric oxide in phytotoxicity-mitigation of copper. Frontiers in Plant Science. 2020;**11**:675

[152] Ahmad P, Latef AAA, Hashem A, Allah EF, Gucel S, Tran L. Nitric oxide mitigates salt stress by regulating levels of osmolytes and antioxidant enzymes in chickpea. Frontiers in Plant Science. 2016;**7**:347

[153] Liu A, Fan J, Gitau MM, Chen L, Fu J. Nitric oxide involvement in Bermuda grass response to salt stress. Journal of the American Society for Horticultural Science. 2016;**141**:425-433

[154] Sehar Z, Masood A, Khan NA. Nitric oxide reverses glucose-mediated photosynthetic repression in wheat (*Triticum aestivum* L.) under salt stress. Environmental and Experimental Botany. 2019;**161**:277-289

[155] Zheng C, Jiang D, Liu F, Dai T, Liu W, Jing Q, et al. Exogenous nitric oxide improves seed germination in wheat against mitochondrial oxidative damage induced by high salinity. Environmental and Experimental Botany. 2009;**67**:222-227

[156] Marvasi M. Potential use and perspectives of nitric oxide donors in agriculture. Journal of Science of Food and Agriculture. 2017;**97**:1065-1072

[157] Kausar F, Shahbaz M, Ashraf M. Protective role of foliar-applied nitric oxide in *Triticum aestivum* under saline stress. Turkish Journal of Botany. 2013;**37**:1155-1165

[158] Fatma M, Khan NA. Nitric oxide protects photosynthetic capacity inhibition by salinity in Indian mustard. Journal of Functional and Environmental Botany. 2014;**4**:106-116

[159] Chen WW, Yang JL, Qin C, Jin CW, Mo JH, Ye T, et al. Nitric oxide acts downstream of auxin to trigger root ferric-chelate reductase activity in response to iron deficiency in *Arabidopsis*. Plant Physiology. 2010;**154**: 810-819

[160] Zhang Y, Wang L, Liu Y, Zhang Q, Wei Q, Zhang W. Nitric oxide enhances salt tolerance in maize seedlings through increasing activities of proton-pump and Na^+/H^+ antiport in the tonoplast. Planta. 2006;**224**:545-555

[161] Wodala B, Deák Z, Vass I, Erdei L, Altorjay I, Horváth F. *In vivo* target sites of nitric oxide in photosynthetic electron transport as studied by chlorophyll fluorescence in pea leaves. Plant Physiology. 2008;**146**:1920-1927

[162] Wu XX, Ding HD, Chen JL, Zhang HJ, Zhu WM. Attenuation of salt-induced changes in photosynthesis by exogenous nitric oxide in tomato (*Lycopersicon esculentum* Mill. L.) seedlings. African Journal of Biotechnology. 2010;**9**:7837-7846

[163] Fatma M, Masood A, Per TS, Khan NA. Nitric oxide alleviates salt stress inhibited photosynthetic performance by interacting with sulfur assimilation in mustard. Frontiers in Plant Science. 2016;7:521

[164] Anjum NA, Umar S, Chan MT. Ascorbate-Glutathione Pathway and Stress Tolerance in Plants. Dordrecht: Springer; 2010

[165] Hasanuzzaman M, Gill SS, Fujita M. Physiological role of nitric oxide in plants grown under adverse environmental conditions. In: Tuteja N, Singh Gill S, editors. Plant Acclimation to Environmental Stress. New York: Springer; 2013b. pp. 269-322

[166] da Silva CJ, Fontes EPB, Modolo LV. Salinity-induced accumulation of endogenous H_2S and NO is associated with modulation of the antioxidant and redox defense systems in *Nicotiana tabacum* L. cv. Havana. Plant Science. 2017;**256**:148-159

[167] Ali Q, Daud MK, Haider MZ, Ali S, Aslam N, Noman A, et al. Seed priming by sodium nitroprusside improves salt tolerance in wheat (*Triticum aestivum* L.) by enhancing physiological and biochemical parameters. Plant Physiology and Biochemistry. 2017;**119**:50-58

[168] Miura K, Tada Y. Regulation of water, salinity, and cold stress responses by salicylic acid. Frontiers in Plant Science. 2014;**5**:4

[169] Khan MIR, Fatma M, Per TS, Anjum NA, Khan NA. Salicylic acid-induced abiotic stress tolerance and underlying mechanisms in plants. Frontiers in Plant Science. 2015;**6**:462

[170] Asensi-Fabado MA, Munné-Bosch S. The aba3-1 mutant of *Arabidopsis thaliana* withstands moderate doses of salt stress by modulating leaf growth and salicylic acid levels. Journal of Plant Growth Regulation. 2011;**30**:456-466

[171] Kaya C, Ashraf M, Alyemeni MN, Ahmad P. The role of endogenous nitric oxide in salicylic acid-induced up-regulation of ascorbate-glutathione cycle involved in salinity tolerance of pepper (*Capsicum annuum* L.) plants. Plant Physiology and Biochemistry. 2020;**147**:10-20

[172] Sousa VF, Santos AS, Sales WS, Silva AJ, Gomes FA, Dias TJ, et al. Exogenous application of salicylic acid induces salinity tolerance in eggplant seedlings. Brazilian Journal of Biology. 2022;**84**:e257739

[173] Quilis J, Peñas G, Messeguer J, Brugidou C, Segundo BS. The Arabidopsis AtNPR1 inversely modulates defense responses against fungal, bacterial, or viral pathogens while conferring hypersensitivity to abiotic stresses in transgenic rice. Molecular Plant-Microbe Interaction. 2008;**21**:1215-1231

[174] Silva KJ, Mahna N, Mou Z, Folta KM. NPR1 as a transgenic crop protection strategy in horticultural species. Horticulture Research. 2018;**5**:15

[175] Yan Y, Pan C, Du Y, Li D, Liu W. Exogenous salicylic acid regulates reactive oxygen species metabolism and ascorbate-glutathione cycle in *Nitraria tangutorum* Bobr. under salinity stress.

Physiology and Molecular Biology of Plants. 2018;**24**:577-589

[176] Csiszár J, Horváth E, Váry Z, Gallé Á, Bela K, Brunner S, et al. Glutathione transferase supergene family in tomato: Salt stress-regulated expression of representative genes from distinct GST classes in plants primed with salicylic acid. Plant Physiology and Biochemistry. 2014;**78**:15-26

[177] Li G, Peng X, Wei L, Kang G. Salicylic acid increases the contents of glutathione and ascorbate and temporally regulates the related gene expression in salt-stressed wheat seedlings. Gene. 2013;**529**:321-325

[178] Jayakannan M, Bose J, Babourina O, Rengel Z, Shabala S. Salicylic acid improves salinity tolerance in Arabidopsis by restoring membrane potential and preventing salt-induced K^+ loss via a GORK channel. Journal of Experimental Botany. 2013;**64**:2255-2268

[179] Alvi AF, Sehar Z, Fatma M, Masood A, Khan NA. Strigolactone: An emerging growth regulator for developing resilience in plants. Plants. 2022;**11**(19):2604

[180] Ma N, Hu C, Wan L, Hu Q, Xiong J, Zhang C. Strigolactones improve plant growth, photosynthesis, and alleviate oxidative stress under salinity in rapeseed (*Brassica napus* L.) by regulating gene expression. Frontiers in Plant Science. 2017;**8**:1671

[181] Ha CV, Leyva-González MA, Osakabe Y, Tran UT, Nishiyama R, Watanabe Y, et al. Positive regulatory role of strigolactone in plant responses to drought and salt stress. Proceedings of the National Academy of Sciences. 2014;**111**(2):851-856

[182] Burger M, Chory J. The many models of strigolactone signaling. Trends in Plant Science. 2020;**25**:395-405

[183] Aroca R, Ruiz-Lozano JM, Zamarreño ÁM, Paz JA, García-Mina JM, Pozo MJ, et al. Arbuscular mycorrhizal symbiosis influences strigolactone production under salinity and alleviates salt stress in lettuce plants. Journal of Plant Physiology. 2013;**170**:47-55

[184] Ren CG, Kong CC, Xie ZH. Role of abscisic acid in strigolactone-induced salt stress tolerance in arbuscular mycorrhizal *Sesbania cannabina* seedlings. BMC Plant Biology. 2018;**18**:74

[185] Sharifi P, Bidabadi SS. Strigolactone could enhances gas exchange through augmented antioxidant defense system in *Salvia nemorosa* L. plants subjected to saline conditions stress. Industrial Crop Production. 2020;**151**:112460

[186] Zhang X, Zhang L, Ma C, Su M, Wang J, et al. Exogenous strigolactones alleviate the photosynthetic inhibition and oxidative damage of cucumber seedlings under salt stress. Scientia Horticulturae. 2022;**297**:110962

Chapter 6

Control of Plant Responses to Salt Stress: Significance of Auxin and Brassinosteroids

Rania Djemal, Moez Hanin and Chantal Ebel

Abstract

Salinity of soils represents a significant abiotic stress factor that not only reduces productivity of most crops but also poses a threat to the global food security. Understanding the mechanisms underpinning plant stress responses as a whole is essential for enhancing crop productivity in salt-affected soils. To improve crop production on salt-affected lands, it is crucial to have a comprehensive understanding of the mechanisms underlying plant stress responses. Phytohormones are key players in these processes, regulating plant growth, development and germination. Among phytohormones, auxin and brassinosteroids (BRs) have been found to overlap to lessen salt stress in plants. In order to help plants balance growth and salt stress tolerance, auxin, BRs, and their interactions are currently known to play a number of important roles. This chapter gives a summary of these findings and discusses how molecular and genetic approaches can be used to engineer auxin, BRs, and thereby develop more salt-resistant cereal crops in the future.

Keywords: salinity, phytohormones, auxin, brassinosteroids, hormonal crosstalk

1. Introduction

Plants cannot avoid the multiple abiotic stresses to which they are constantly exposed because of their sessile nature. Stresses like salinity, drought, cold and heat greatly hinder the growth and productivity of plants. Among those, soil salinity significantly affects crop output and growth all over the world [1]. It is estimated that by 2050, roughly 20% of agricultural land will be unproductive due to soil salinization, which will affect nearly 50% of agricultural land [2]. Salt stress occurs when there is too much Na^+ in the soil solution, preventing plants from receiving water and nutrients from the soil. Na^+ accumulation is harmful because it causes osmotic and ionic stresses that promote the formation of ROS, alter plant metabolism, and upset the balance of ions in the environment [3].

Based on their capacity to flourish in salty conditions, plants are divided into two main groups called glycophytes and halophytes. Since most cultivated plants are glycophytes, they cannot withstand salty environments with concentrations of more than 100 mM NaCl. In the case of cereals, rice (*Oryza sativa*) is classified as the least tolerant species, followed by durum wheat (*Triticum durum*), common wheat (*Triticum aestivum*), maize (*Zea mays*) and barley (*Hordeum vulgare*) which is considered as the most tolerant species [4].

A proper hormonal balance is required in plants in order to limit the potential negative impacts of environmental variables. Most phytohormones are know to play a major role in controlling plant growth and development as well as stress responses [5–8]. Thanks to their extensive range of functions and complex interplay, auxin and brassinosteroids (BRs) are considered as two of the phytohormones that hold the most promise for tailoring abiotic stress tolerance in crop plants [9, 10]. Many auxin-responsive genes have been demonstrated to be synergistically regulated by the interaction between BRs and auxin pathways [11]. Furthermore, through separate processes, auxin and BRs can stimulate root development and increase cell expansion [12]. This chapter attempts to offer new insights into our understanding of how auxin, BR and their interactions can support plants' ability to balance growth and salt stress tolerance (**Figure 1**).

2. Auxins: major roles in plants under salinity stress

Many physiological and developmental processes, including the formation of lateral and adventitious roots, flowering, senescence, and morphogenesis, are regulated by auxins, primarily indole-3 acetic acid (IAA) [13]. Auxin is one of the most significant phytohormones involved in the regulation of lateral root growth, main root elongation, and halotropism, the special capacity of plants to avoid salty circumstances when they are under salt stress. Auxin-mediated lateral root formation and the cessation of their growth in response to excessive salt have been shown to be antagonistic [14]. The gradient, concentration, and spatiotemporal expression of receptor genes tightly govern auxin's regulatory mechanisms [15]. The intense regulation of this phytohormone at several levels, as well as its manufacture and signaling, led to a drop in endogenous auxin levels being identified under salt stress [16]. Three key auxin sources exist in plants, namely synthetic auxin, endogenous auxin and microbial auxin sources from the rhizosphere [17].

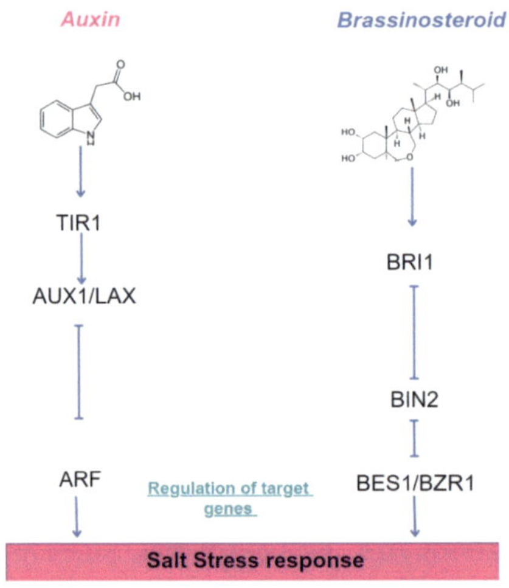

Figure 1.
Scheme representing the major structure of auxin and brassinosteroids, and their connection with salt stress.

Auxin signal transduction pathway has been broadly examined [18, 19] and TIR1 encodes a nuclear auxin receptor belonging to the F-box protein [20] that interacts with a group of AUX/IAA (Auxin/Indole-3-Acetic Acid) proteins. Through their interaction with the transcriptional regulators of ARF (auxin response factors), AUX/IAA operates as negative regulators by impeding transcriptional auxin output [21]. ARFs, with their 23 members, can therefore activate or repress target genes and mediate the auxin responses, and they are destroyed when auxin binds to TIR1 [22].

Several kinds of efflux/influx carriers control the auxin transport in the plant root, resulting in the auxin gradient. The auxin-resistant 1/like aux1 (AUX/LAX) family of influx carriers mediates an active polar transport [23]. On the other hand, auxin efflux transporters include ABC transporter family [24], NRT1/PRT family of nitrate transporter 1/peptide transporter [25], and PIN-FORMED carriers [26, 27]. PINs constitute a family of 7 proteins that are found in the plasma membrane and are thought to be involved in the control of auxin transport [17]. Salt stress lowers auxin levels, which in turn drops the expression of auxin transporters [16], linking auxin distribution and biosynthesis [28]. PIN abundance is primarily responsible for the disruption in auxin transport. The authors in [29] have discovered that the PIN2 auxin efflux carrier is the most specific to salt stress since it is seen to actively redistribute auxin in the root tip when exposed to a salt gradient. Auxin redistribution and directional bending of the root away from high salt levels are mediated by PIN2 internalization, which is stimulated by salt-induced phospholipase-D at the side of the root facing the higher salt concentration. Under salt stress, the downregulation of PIN1, PIN3 and PIN7 is also an important part of the asymmetric distribution of auxin, which affects root bending from the salt [15]. Furthermore, primary root size is abridged under salt stress alongside with a decrease in lateral root density due to decreased levels of PIN1, PIN3 and PIN7 [16].

Auxin carriers' function can be regulated through post-translational changes, subcellular localization, and regulation of their expression in addition to the regulation of their expression. In order to allow auxin redistribution and for the directional bending of the root away from the higher salt concentration, PIN2 and AUX1 alter their subcellular location in endosomes [29, 30]. AUX1 and PIN2 are required for the establishment of gravity-inducing asymmetric auxin response. Auxin transport in the elongation zone requires AUX1, but its transport back to the root tip is mostly mediated by PIN2 [31]. According to the model of [32] that was predicted to occur during halotropism, auxin asymmetry is caused by an imbalance in the PIN2 and AUX1 pathways in the root tip, with PIN2 decreasing on the side of the root that is exposed to salt and changing auxin levels on the opposite side. This asymmetry of auxin was amplified by the AUX1 auxin transporter. The AUX1 auxin transporter increased this auxin asymmetry.

What is trusty to mention is that PIN polarity and intracellular auxin polar fluxes are both required for PIN phosphorylation [33]. For instance, during phototropic and gravitropic reactions mediates PIN3 phosphorylation to establish an auxin gradient during phototropic and gravitropic responses [34]. The balanced actions of PID kinase and protein phosphatases do in fact regulate the status of phosphorylation in PINs. RCN1 (ROOTS CURL IN NPA 1) encodes a regulatory subunit of protein phosphatase 2A (PP2A). It has been reported that the mutant rcn1 shows an elevation in gravitropic root bending curvature [33–35]. Auxin biosynthesis is carried out by different reactions of the indole-3-acetaldoxime pathway mediated by enzymes like tryptophan aminotransferase (TAA)/YUCCA (YUC) [36]. The processes of IAA production, conjugation, and degradation determine the amounts of IAA in cells. Firstly, the YUCCA (YUC) family of enzymes, which are present throughout the root system, controls IAA production [37]. Auxin redistribution caused by the movement of auxin production from columella cells to the root epidermis during salt stress is partially connected with a reduction in the growth of primary and lateral roots [38]. Interestingly, YUC gene expression is calibrated in response to salt stress [39]. In fact, transcriptomic data show that YUC5 is up-regulated immediately after salt exposure [40] demonstrating that this gene plays a major role in auxin-mediated salt stress response.

Secondly, IAA levels are impacted by both conjugation and degradation processes, as observed through the positive correlation between free IAA levels and the levels of IAA conjugates and catabolites in roots and shoots. The degradation of IAA is the primary factor responsible for its rapid turnover and is mainly catalyzed by DAO1 and 2. A study has shown that DAO1/2 plays a crucial role in the root response to salt stress induced by lateral root density. Indole-3-butyric acid (IBA) contributes significantly to lateral root growth, root hair elongation, and adventitious root formation during root development. However, IBA can undergo oxidation and transform into IAA, which can negatively impact a plant's ability to tolerate salinity [41]. The irreversible conjugation of aspartic acid to IAA tags it for oxidation as well as for catabolism and IAA-asp conjugates are known to be involved in IAA detoxification [42].

It has been established that abiotic stress, particularly salt, has an impact on these processes. The GH3 (Gretchen Hagen) family of enzymes can catalyze the addition of some groups and generate diverse auxin conjugate genes (i.e: ILR/IAR (IAA-amido hydrolase) and ILL (ILR1 likes)). Salt stress tolerance is mediated in part by auxin homeostasis, which is controlled by a collection of GH3 enzymes through a negative feedback regulation. WES1 (a kind of GH3) gene regulation is essential for salt stress tolerance [43]. Indeed, GH3 gene family is a desirable option for salt stress breeding since it is extensively expressed in roots and is increased when exposed to salt stress [39].

3. Brassinosteroids: major roles in plants under salt stress

Brassinosteroids (BRs) are a group of plant steroid hormones, firstly isolated from Brassica pollen. There are about 60 compounds [11, 44, 45] of which the most bioactive BRs are brassinolide, 24-epibrassinolide, and 28-homobrassinolide. The identification of BR signaling components through molecular and genetic studies has provided a significant breakthrough in biotechnological modification for enhancing crop yield and stress tolerance in the context of global climate change. This pathway is considered as the most crucial target for these modifications [46–49]. Recent research has confirmed the significant role of BR in the regulation of various physiological processes in plants, such as the regulation of metabolic reactions in response to different biotic and abiotic conditions. This includes responses to pathogen-triggered reactions, as well as the management of salt and drought stress, scavenging of Reactive Oxygen Species (ROS), and reactions to herbicides and pesticides [49]. In addition to its regulatory role in plant physiology, BR also plays a crucial role in morphogenetic processes during plant growth and development. However, this regulation is complex and involves a sophisticated interplay between the components of the BR signaling pathway and the signal transduction pathways of other phytohormones [50].

The perception of BRs begins with the binding of the hormone to a transmembrane polypeptide BRI1 (Brassinosteroid-Insensitive 1) as well as its co-receptor BAK1

(BRASSINOSTEROID INSENSITIVE 1-associated receptor kinase 1) both of which belong to Leucine-Rich repeat Receptor-like kinases (LRR-RLK 1) family [51]. After the perception of BR by BRI1, the signal is transduced by several events of reversible phosphorylation leading to the activation of a BRASSINAZOLE RESISTANT 1 (BZR1), and BZR2, also known as BRI1-EMS SUPPRESSOR1 (BES1) transcription factors [52]. After being perceived by BRI1 and BAK1, unphosphorylated BZR1 and BES1 move into the nucleus and regulate the expression of their target genes [53–56]. This pathway is highly complex and involves the interplay of numerous proteins, including PP2A, which regulates both BRI1 and BES1 in the pathway. Furthermore, recent studies have highlighted the contribution of a type 1 protein phosphatase (TdPP1) in the dephosphorylation of BES1 and its subsequent activation upon BR treatment. These findings have shed new light on the intricacies of the BR signaling pathway and its potential for further biotechnological advancements in crop yield and stress tolerance.

Decades of research have explored the relationship between BR and plant stress response. One study [57] reported that treating barley with BR improved salt tolerance, potentially through the regulation of water loss via reduced stomatal conductance and density [58]. In many species, BR application through the root-growing media has been shown to promote seed germination under salt stress. In *Brassica napus*, the inhibitory effect of salt stress was reduced by the addition of BRs to the germination medium [59]. Another study [60] has demonstrated that pre-soaking rice seeds with BRs and NaCl alleviates the inhibitory effect of salt on seed germination and seedling growth associated with increased levels of nucleic acids and soluble proteins in the kernel of rice.

Maize plants treated with BRs have been shown to alleviate oxidative stress in salt, leading to improved seedling growth and reduced lipid peroxidation, likely through the induction of antioxidant enzyme activities such as CAT, SOD and POD [61]. Similarly, in wheat, the addition of exogenous BRs has been shown to enhance plant growth under saline conditions [62]. In Arabidopsis, endogenous BR is positively involved in the plant response to salt stress as confirmed by the hypersensitivity of BR-deficient mutant det2-1 and BR-insensitive mutant bin2-1 to salt stress during seedling growth and seed germination. This hypersensitivity is correlated with the inhibited induction of stress-related genes, namely P5CS1, COR78 and proline accumulation under salt stress conditions [63]. Furthermore, the addition of exogenous BR has been shown to improve NaCl-induced proline accumulation and eliminate the inhibition of root elongation in WT plants [63]. These findings provide insight into the potential of BR-based biotechnological interventions for enhancing plant stress response and improving crop yield under adverse conditions.

4. Auxin-brassinosteroids crosstalk: an important approach for plant salt stress tolerance

The interplay between auxin and BRs is known to regulate various aspects of plant development and growth, not just at the individual level but also in cross-talk [64]. Despite being a well-investigated concept for over a decade, it is not yet fully understood, particularly in crops. Studies have shown that the root level auxin and BR exhibit opposed actions, with an optimal expression of BZR1 depending on auxin biosynthesis [65]. BR catabolism and BR-mediated signaling lead to the specific spatiotemporal activation of auxin-related genes in the elongation zone, while repressing them in the quiescent center [64]. Indeed, BZR1 directly interacts with ARF proteins to target multiple auxin-related genes, including those involved in transport and signaling, such as AUX/IAA, PINs and TIR/AFB. Therefore, ARFs genes are composed of a carboxy-terminal dimerization domain that facilitates protein-protein interactions, not only within the AUX/IAA family but also between ARF genes. [66]. In addition, it has been established that BR signaling connects with SOB3 (SUPPRESSOR OF PHYTOCHROME B4-3) to control cell elongation and hypocotyl growth through the up-regulation of SAUR19 (SMALL AUXIN UP RNA19) expression [67].

BR plays a crucial role in the transport of auxin by affecting the cellular localization of auxin efflux and influx carriers such as PIN3, PIN4 and AUX1/LAXs [12–74]. Specifically, BR controls accumulation of intracellular auxin flow PIN2 from the root tip towards the shoot by recycling it back to the vasculature via the lateral root cap and epidermis. The accumulation of PIN2 and PIN4 is regulated by BR in a post-transcriptional manner, and BR has a similar effect on PIN21 and PIN4 accumulation in collumella [74]. During plant gravitropism, BR intensifies the accumulation of the PIN2 gene in the root meristem zone and affects the allocation of auxin from the root tip towards the elongation zones, resulting in a difference in IAA levels in the upper and lower sides of roots. It has been demonstrated that during this process, BR activates ROP2 which plays a vital role in modulating the functional localisation of PIN2 through the regulation of F-actins. In contrast, BRX (brevis radix) which regulates cell proliferation and elongation in the roots and shoots is vastly brought by auxin and repressed by BRs [11]. Interestingly, the BR-biosynthetic genes, DWF4 and CPD, have been shown to be activated BRX, highlighting the functional relationship between auxin signaling and BR biosynthesis [75]. The connection between auxin and BR is also evident when roots are treated with exogenous auxin, which increases DWF4 expression, leading to an increase in BR biosynthesis. However, when BR is synthesized, DWF4 is retro-inhibited by BR itself [76]. Several studies have suggested that the effects of BR are also influenced by auxin, either by enhancing sensitivity to this hormone or by altering its levels [77, 78]. According to a study by [79], GH3 genes in soybean and tomato were not promptly activated during BR-induced cell expansion but were activated by BR after cell elongation had begun.

Auxin and BR have a synergistic relationship that is evident in their combined effects on root development. One example of this interaction is demonstrated through the interplay between BIN2 and ARF2 repressors. [76]. BIN2 was found to phosphorylate ARF2, which inhibits its interaction with the AUX/IAA repressor and enhances auxin response [80]. ARF2 is also a target of BZR1 and its expression is decreased by BR treatment [54]. Phosphorylation by BIN2 can reach additional ARFs (ARF7 and ARF19) to induce the transcriptional activity of their target genes LATERAL ORGAN BOUNDARIES-DOMAIN16 (LBD16) and LBD29 acting on lateral root organogenesis [81]. In Arabidopsis, many regulators genes are known to control seed size and endosperm development like SHB1 (SHORT HYPOCOTYL UNDER BLUE 1), IKU1 (HAIKU 1), IKU2 (HAIKU 2), and MINI3 (MINISEED 3) through their interaction in BR-signaling [82]. These proteins are involved in the regulation of BZR1 under the control of the BR-BRI1-BIN2 phosphorylation cascade. Therefore, BR can inhibit the expression of APETALA 2 (AP2), the floral homeotic gene, and AUXIN RESPONSE FACTOR 2 (ARF2), the key negative regulators of seed size and weight [83]. Moreover, the exogenous application of BR induced the expression of auxin-responsive genes implicated in root development of which we can cite IAA7, IAA17 and IAA14. However, BR signaling mutant and biosynthetic mutant det2 and bri1 had significantly decreased gene expression of AXR3/IAA17 as well as several

Aux/IAA genes, such as AXR2/IAA7, SLR/IAA14, and IAA28. This finding suggests that BR signaling pathways and auxin signaling pathways are integrated during root development [84]. The interaction between BR and auxin is also involved in regulating plant stress responses. In cucumber plants subjected to different stress conditions, including salt, cold, and PEG, the expression of many YUCCA genes is reduced. However, yucca mutants exhibit higher levels of transcripts of BR-related genes such as BRI1 [85].

Despite the importance of BRs and auxin in regulating salt stress response and root growth, the specific molecular mechanisms involved in this process are still unclear. It has been observed that, under salt stress conditions, transcription factors associated with the BR pathway can affect auxin homeostasis by modulating the expression of genes involved in auxin biosynthesis, conjugation, and degradation. (**Figure 2**; **Table 1**). On the other hand, BZR1/BES1 dephosphorylation causes the rapid induction of genes encoding the auxin biosynthetic enzymes like YUC7/3/8/5 under salt stress. Thus, auxin and BRs signaling participate in regulating a large spectrum of root developmental processes by the formation of an auxin gradient, allowing plant seedlings to cope with salinity. This movement of local auxin concentration was regulated by the expression of CYP79B2, ABCB family, PIN, YUC, GH3 and PAT1 (PHOSPHORIBOSYL ANTHRANILATE TRANSFERASE) especially under abiotic stress by heavy metal [117]. Gene expression studies have revealed that genes involved in tryptophan-dependent IAA biosynthesis pathway like YUC4, NIT1; NIT2, and IAA degradation like DAO were increased by salt stress [118]. Transcriptomic data indicated that some IAA biosynthesis genes such as AAO1 (ARABIDOPSIS ALDEHYDE OXIDASE1); CYP79B2,3 (CYTOCHROME P450 FAMILY 79B2,3) and AMI1 (INDOLE-3-ACETAMIDE) display similar expression pattern under salt stress and control conditions. Nonetheless, the expression of DAO (DIOXYGENASE FOR

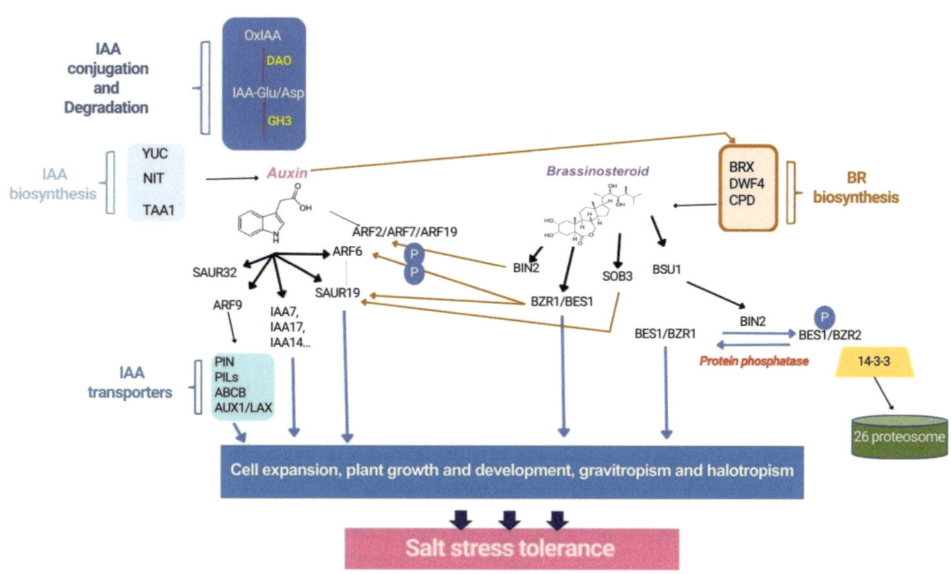

Figure 2.
A schematic model highlighting the potential molecular-genetic mechanisms involved in the auxin-brassinosteroids crosstalk under salt stress. Black solid arrows show regulation, blue solid arrows salt stress tolerance via IAA and BR signaling pathway. Red solid arrows show crosstalk between IAA and BR signaling pathway.

Genes involved in the crosstalk BR-auxin	Gene accession number	Description	Ref.
YUC (YUCCA)	AT4G32540	Catalyzes the conversion of IPA (indole-3-pyruvic acid) to IAA	[86]
NIT1 (NITRILASE1)	AT3G44310	Participates in the conversion of IAN (indole-3-acetonitrile) to IAA	[86]
NIT2(NITRILASE2)	AT3G44300	Participates in the conversion of IAN (indole-3-acetonitrile) to IAA	[86]
TAA1 (TRYPTOPHAN AMINOTRANSFERASE)	AT1G70560	Encodes an aminotransferase that converts Trp (Tryptophane) to IPA (, indole-3-pyruvic acid)	[87, 88]
SAUR (SMALL AUXIN UPREGULATED RNA)	AT1G11803	Encode Small Auxin Up RNAs, that regulate leaf growth through controlling cell expansion or division, contributing to auxin-regulated leaf growth and development.	[89, 90]
AXR1 (AUXIN RESISTANT1)	AT1G05180	Encode a protein related to the ubiquitin-activating enzyme, and its function in the ubiquitin conjugation pathway	[91]
GH3 (GRETCHEN HAGEN3)	AT1G48660	Encodes an auxin-conjugating enzyme; and contributes to the auxin-mediated modulation of plant growth in response to environmental stresses.	[92]
PIN1 (PIN-FORMED 1)	AT1G73590	Encodes an auxin efflux carrier involved in shoot and root development. It is involved in the maintenance of embryonic auxin gradients.	[16, 93]
PIN2 (PIN-FORMED 2)	AT5G57090	PIN2 functions as an auxin efflux facilitator mediating proximal shootward auxin transport in the Arabidopsis root	[94]
PIN3 (PIN-FORMED 3)	AT1G70940	PIN3 is an auxin efflux carrier which is expressed in the cortical cells situated in front of the LRP(Lateral root primordium) and it is induced by auxin in this tissue.	[16, 95]
PIN4 (PIN-FORMED 4)	AT2G01420	PIN4 is involved in the regulation of auxin homeostasis and patterning through sink-mediated auxin distrubition in root tips.	[16, 96]
PIN5 (PIN-FORMED 5)	AT5G16530	PIN5 regulates intracellular auxin homeostasis and metabolism.	[16, 96]
PIN6 (PIN-FORMED 6)	AT1G77110	PIN6 is an important component of auxin transport and auxin homeostasis and contributes to auxin-dependant growth and development processes such as root and shoot.	[16, [97]
PIN7 (PIN-FORMED 7)	AT1G23080	PIN7 involved in the redistribution of auxin from the maximum in the collumella initials to the epidermis and lateral root cap. Gravity stimulation of roots induces rapid polarization of PIN7 towards the lateral plasmamembrane.	[16, 98]
SAUR32 (SMALL AUXIN UPREGULATED RNA 32)	AT2G46690	SAUR32 is implicated in the reduction of hypocotyl growth and abolished apical hook formation in the dark.	[39, 99]

Genes involved in the crosstalk BR-auxin	Gene accession number	Description	Ref.
SAUR19 (SMALL AUXIN UPREGULATED RNA 19)	AT5G18010	SAUR19 is considered as a positive effector of cell expansion. The regulation of this gene is achived through the modulation of auxin transport.	[39, 100]
DAO1(DIOXYGENASE FOR AUXIN OXIDATION 1)	AT1G14130	DAO1 catalyzes the formation of oxIAA and regulate auxin homeostasis and plant growth. DAO1 plays a crucial role in plant morphogenesis.	[101]
DAO2 (DIOXYGENASE FOR AUXIN OXIDATION 2)	AT1G14120	DAO2 catalyzes the oxidation of IAA into oxIAA. DAO2 is expressed in root tip.	[101]
ARF7 (AUXIN RESPONSE FACTOR 7)	AT5G20730	ARF7 is considered as a transcriptional activators of auxin-response genes. ARF7 Regulates Lateral Root Formation via Direct Activation of LBD/ASL Genes in Arabidopsis	[102]
ARF19 (AUXIN RESPONSE FACTOR 19)	AT1G19220	ARF19 Regulate Lateral Root Formation via Direct Activation of LBD/ASL Genes in Arabidopsis.	[102]
ARF2 (AUXIN RESPONSE FACTOR 2)	AT5G62000	The main function of ARF2 is in the auxin-mediated control of Arabidopsis leaf longivety.	[103]
ARF9 (AUXIN RESPONSE FACTOR 9)	AT4G23980	ARF9 reveals en inhibitory effect on auxin-responsive root hair growth.	[104, 105]
AXR2/IAA7 (AUXIN RESISTANT2/INDOLE-3-ACETIC ACID 7)	AT3G23050	AXR2/IAA7 is one of the core components of the auxin signaling pathway and it is involved in stem development.	[106, 107]
AXR3/IAA17 (AUXIN RESISTANT3/INDOLE-3-ACETIC ACID 17)	AT1G04250	The involvement of AXR3/IAA17 gene expression in brassinosteroid (BR)-regulated root development. It is involved in the reduction of root elongation and cause and increased adventitious root.	[16, 84]
IAA14 (INDOLE-3-ACETIC ACID 14)	AT4G14550	IAA14 is a transcriptional repressor of auxin signaling.	[108, 109]
AUX1/LAX1 (AUXIN RESISTANT 1/LIKE AUX1)	AT5G01240	AUX1/LAX is an auxin influx carriers, AUX1/LAX genes are implicated in the regulation of key plant processing including root, lateral root development, root gravitropism, root hair development and leaf morphogenesis.	[110, 111]
ABCB (ATP-BINDING CASETTESUBFAMILY B)	AT1G02520	ABCB participates in polar movement of auxin by exclusion from and prevention of uptake at the plasma membrane.	[110, 111]
PILS1 (PIN-LIKES1)	AT1G20925	PILS proteins are putative auxin carriers that regulate the auxin transport from the cytosol into the lumen of the ER and also it mediate intracellular auxin accumulation	[112]
DWF4 (DWARF4)	AT3G50660	DWF4 is involved in BR Biosynthesis	[113]
CPD (l-(2'-CARBOXYPHENYL)-3-PHENYLPROPANE-1,3-DIONE)	AT5G05690	CPD is a BR-specific biosynthesis genes	[114]

Genes involved in the crosstalk BR-auxin	Gene accession number	Description	Ref.
BZR1/BES1 (BRASSINAZOLE-RESISTANT 1/BRI1-EMS-SUPPRESSOR)	AT1G75080	BES1 and BZR1 can directly or indirectly regulate the expression of thousands of BR-responsive genes and ultimately affect plant growth, development, and stress adaptation	[115]
BIN2 (BRASSINOSTEROID INSENSITIVE2)	AT4G18710	BIN2 encodes a negative regulator of BR-signaling in plant growth.	[67]
SOB3 (SUPPRESSOR OF PHYB)	AT1G76500	SOB3 was present in BR-signaling and it is implicated in the transcription of genes envolved in cell elongation and hypocotyl growth.	[67]
BSU1 (BRI1 SUPPRESSOR 1)	AT1G08420	BSU1 dephosphorylates and inactivates downstream BRASSINOSTEROID INSENSITIVE2	[116]

Table 1.
The most important genes involved in the crosstalk between BRs and auxin.

AUXIN DEGRADATION), which is involved in auxin degradation, did not change significantly in plants grown in the presence of NaCl [118]. Both BR and auxin are recognized as key regulators that exert gradual effects on a range of growth processes, including cell division and cell elongation, particularly under abiotic stress conditions [65]. Recent research has revealed that ARF and BZR collaborate to promote hypocotyl elongation [119]. Similarly, BRs activate the expression of SAUR19 via BZR1 and there is evidence of interaction between ARF6, BZR1 and SAUR genes [67]. More recently, it has been demonstrated that these genes are implicated in enhancing plant tolerance to abiotic stress, particularly drought stress [120].

5. Auxin-Brassinosteroids pathways: molecular and genetic perspectives for improved salt tolerance in cereal crops

Hormone signaling and metabolism pathways are widely regarded as promising targets for improving abiotic stress tolerance in plants, particularly in cereal crops [5]. As a result, maintaining the balance of phytohormones is critical for promoting optimal growth and development in plants [121]. Auxin and brassinosteroid are among the most commonly investigated phytohormones for improving abiotic stress tolerance in crops. These hormones are known to play a crucial role in mitigating salt stress, and they exhibit a diverse range of functions in this regard [122]. Consequently, several key enzymes involved in auxin and brassinosteroid signaling pathways have been genetically engineered to enhance abiotic stress tolerance in plants [121]. However, limited information is available on the mechanisms underlying the crosstalk between brassinosteroids and auxin in cereal crops. Nonetheless, transferring knowledge from model plant species such as Arabidopsis to cereal crops like wheat and barley is advantageous, as the signaling mechanisms in plants are evolutionarily conserved across [49]. Regarding Br signaling [123], discovered a negative correlation between the brassinosteroid pathway and abiotic stress tolerance in *Brachypodium distachyon*. In particular, when the mutant form of the BRI1 gene (bri1) was present, the plant exhibited an improvement in abiotic stress tolerance, particularly under drought stress conditions. Furthermore, in rice BZR1, a dependent BR-gene, functions as a positive regulator of the BR signaling. The RNAi-mediated silencing of the OsBZR1 gene expression results in the BR insensitivity, semi-dwarfism and erect phenotype [124].

In rice, Hwang et al. [125] have demonstrated that a complex composed of BR-OsBRI1-OsBAK1 inactivates the OsGSK2 which, in turn, inactivates the BR signaling output regulators, namely OsBZR1, LIC (TILLER ANGLE INCREASED CONTROLLER), OsGRF4 (Growth-Regulating Factor), and CYC-U2 (cyclin U-type) in rice. De-phosphorylated OsBZR1 regulates the target components (CYC-U4;1, LIC, ILI (lilliputian1), and DLT (LOW-TILLERING)) involved in primary BR response in rice. Another BR-signaling component, SERK2, was identified in rice cultivars. The generation of mutant alleles of SERK2 by CRISPR/Cas9 editing showed a higher sensitivity to salt stress with an increase of grain size. In contrast, the overexpression of SERK2 enhances resistance of plants to salt stress without affecting plant architecture [126].

Therefore, plants subjected to abiotic stress especially salinity and exogenous BR exhibit two main patterns of gene regulation: (i) BR rescue expression of developmental proteins that are suppressed under salt stress and (ii) BR induce higher levels of protective proteins than salt stress alone. Transcriptomic analyses reveal that salt stress causes the downregulation of many genes critical to cell wall synthesis, photosynthesis carbon assimilatory process, starch transport and accumulation, as well as many metabolic pathways [127]. Contrariwise, BR up-regulated genes are associated with plant growth and development processes, targeting genes encoding cell elongation and cell wall modification enzymes, auxin responsive factors, and TFs, among others, indicating the mechanisms by which BR act to alleviate abiotic stress especially salt stress [47]. A recent study identified a specific interaction between BR and auxin pathways RLA1/SMOS1 (REDUCED LEAF ANGLE 1/SMALL ORGAN SIZE 1), a transcriptional regulator of BR signaling pathway, which form a complex with OsBZR1 and activates the BR signal transduction [125]. In rice, it has been shown that the RLA1/SMOS1 gene can be activated by auxin, indicating a potential crosstalk between the auxin and brassinosteroid pathways [128]. In addition to the previously mentioned RLA1/SMOS1 genes, LPA1 (Loose Plant Architecture1) has also been found to play a role in regulating plant architecture and auxin homeostasis in rice [129]. Two BR-mediated pathways are two BR-mediated pathways that interact with auxin to regulate the leaf inclination in rice: the BR biosynthesis-dependent pathway and the OsBRI1-mediated pathway. LPA1 has been shown to inhibit auxin signaling by interacting with C-22-hydroxylated and 6-deoxo BRs, independently of the OsBRI1-mediated pathway. However, there is no evidence of a direct interaction between OsBRI1 and LPA1 proteins [130].

6. Conclusions and prospects

Recently, advanced research has focused on unraveling the genetic and molecular mechanisms of BR and auxin pathways in plants responses to salt stress. BR plays a crucial role in promoting responses to salt stress, by activating a family of transcription factors BZR1/BES1 which can directly or indirectly regulate the expression of BR-responsive genes. It can ultimately affect not only development and growth of plants but also auxin biosynthesis and signaling. The interplay between auxin and BR pathways has been shown to be crucial in multiple plant development processes under salt stress including hypocotyl elongation, root development, halotropism and gravitropism. Future thorough studies are required to fully understand the interdependency between auxin and BR in improving salt stress tolerance before implementing new approaches to engineer stress resilient crops.

Author details

Rania Djemal*, Moez Hanin and Chantal Ebel
Plant Physiology and Functional Genomics Research Unit, Institute of Biotechnology of Sfax, University of Sfax, Sfax, Tunisia

*Address all correspondence to: raniajmalabid86@gmail.com

References

[1] Shah T, Latif S, Saeed F, Ali I, Ullah S, Abdullah Alsahli A, et al. Seed priming with titanium dioxide nanoparticles enhances seed vigor, leaf water status, and antioxidant enzyme activities in maize (*Zea mays* L.) under salinity stress. Journal of King Saud University - Science. 2021;**33**(1):101207. DOI: 10.1016/j.jksus.2020.10.004

[2] Li X, Xu S, Fuhrmann-Aoyagi MB, Yuan S, Iwama T, Kobayashi M, et al. CRISPR/Cas9 technique for temperature, drought, and salinity stress responses. Current Issues in Molecular Biology. 2022;**44**(6):2664-2682

[3] Kundu P, Gill R, Ahlawat S, Anjum NA, Sharma KK, Ansari AA, et al. Targeting the redox regulatory mechanisms for abiotic stress tolerance in crops. In: Biochemical, Physiological and Molecular Avenues for Combating Abiotic Stress in Plants. Elsevier Inc.; 2018. pp. 151-220. DOI: 10.1016/B978-0-12-813066-7.00010-3

[4] Munns R, James RA, Läuchli A. Approaches to increasing the salt tolerance of wheat and other cereals. Journal of Experimental Botany. 2006;**57**(5):1025-1043

[5] Sreenivasulu N, Harshavardhan VT, Govind G, Seiler C, Kohli A. Contrapuntal role of ABA: Does it mediate stress tolerance or plant growth retardation under long-term drought stress? Gene. 2012;**506**(2):265-273. DOI: 10.1016/j.gene.2012.06.076

[6] Cortleven A, Leuendorf JE, Frank M, Pezzetta D, Bolt S, Schmülling T. Cytokinin action in response to abiotic and biotic stresses in plants. Plant, Cell & Environment. 2019;**42**(3):998-1018

[7] Ku YS, Sintaha M, Cheung MY, Lam HM. Plant hormone signaling crosstalks between biotic and abiotic stress responses. International Journal of Molecular Sciences. 2018;**19**:3206-3241

[8] Peleg Z, Blumwald E. Hormone balance and abiotic stress tolerance in crop plants. Current Opinion in Plant Biology. 2011;**14**(3):290-295. DOI: 10.1016/j.pbi.2011.02.001

[9] Wei H, Jing Y, Zhang L, Kong D. Phytohormones and their crosstalk in regulating stomatal development and patterning. Journal of Experimental Botany. 2021;**72**(7):2356-2370

[10] Chaudhuri A, Halder K, Abdin MZ, Majee M, Datta A. Abiotic stress tolerance in plants: Brassinosteroids navigate competently. International Journal of Molecular Sciences. 2022;**23**(23):14577-14594

[11] Mouchel CF, Osmont KS, Hardtke CS. BRX mediates feedback between brassinosteroid levels and auxin signalling in root growth. Nature. 2006;**443**(7110):458-461

[12] Bao F, Shen J, Brady SR, Muday GK, Asami T, Yang Z. Brassinosteroids interact with auxin to promote lateral root development in arabidopsis 1. Plant Physiology. 2004;**134**(4):1624-1631

[13] Okushima Y, Mitina I, Quach HL, Theologis A. AUXIN RESPONSE FACTOR 2 (ARF2): A pleiotropic developmental regulator. The Plant Journal. 2005;**43**(1):29-46

[14] Julkowska MM, Testerink C. Tuning plant signaling and growth to survive salt. Trends in Plant Science. 2015;**20**(9):586-594. DOI: 10.1016/j.tplants.2015.06.008

[15] van den Berg T, ten Tusscher KH. Auxin information processing; partners

and interactions beyond the usual suspects. International Journal of Molecular Sciences. 2017;**18**(12):2585-2599

[16] Liu W, Li RJ, Han TT, Cai W, Fu ZW, Lu YT. Salt stress reduces root meristem size by nitric oxidemediated modulation of auxin accumulation and signaling in Arabidopsis. Plant Physiology. 2015;**168**(1):343-356

[17] Ribba T, Garrido-Vargas F, Antonio O'brien J. Auxin-mediated responses under salt stress: From developmental regulation to biotechnological applications. Journal of Experimental Botany. 2020;**71**(13):3843-3853. Available from: https://academic.oup.com/jxb/article/71/13/3843/5841204

[18] Yu Z, Zhang F, Friml J, Ding Z. Auxin signaling: Research advances over the past 30 years. Journal of Integrative Plant Biology. 2022;**64**(2):371-392

[19] Leyser O. Auxin Signaling. Vol. 176. Plant Physiology. American Society of Plant Biologists; 2018. pp. 465-479

[20] Dharmasiri N, Dharmasiri S, Estelle M. The F-box protein TIR1 is an auxin receptor. Nature. 2005;**435**(7041):441-445

[21] Lanza M, Garcia-Ponce B, Castrillo G, Catarecha P, Sauer M, Rodriguez-Serrano M, et al. Role of actin cytoskeleton in Brassinosteroid signaling and in its integration with the auxin response in Plants. Developmental Cell. 2012;**22**(6):1275-1285

[22] Guilfoyle TJ, Hagen G. Auxin response factors. Current Opinion in Plant Biology. 2007;**10**(5):453-460

[23] Swarup R, Péret B. AUX/LAX family of auxin influx carriers-an overview. Frontiers in Plant Science. 2012;**3**(OCT):1-11

[24] Okamoto K, Ueda H, Shimada T, Tamura K, Koumoto Y, Tasaka M, et al. An ABC transporter B family protein, ABCB19, is required for cytoplasmic streaming and gravitropism of the inflorescence stems. Plant Signaling & Behavior. 2016;**11**:3

[25] Chiba Y, Shimizu T, Miyakawa S, Kanno Y, Koshiba T, Kamiya Y, et al. Identification of Arabidopsis thaliana NRT1/PTR FAMILY (NPF) proteins capable of transporting plant hormones. Journal of Plant Research. 2015;**128**(4):679-686. DOI: 10.1007/s10265-015-0710-2

[26] Habets MEJ, Offringa R. PIN-driven polar auxin transport in plant developmental plasticity: A key target for environmental and endogenous signals. The New Phytologist. 2014;**203**(2):362-377

[27] Barbez E, Kubeš M, Rolčík J, Béziat C, Pěnčík A, Wang B, et al. A novel putative auxin carrier family regulates intracellular auxin homeostasis in plants. Nature. 2012;**485**(7396):119-122

[28] Shen CJ, Bai YH, Wang SK, Zhang SN, Wu YR, Chen M, et al. Expression profile of PIN, AUX/LAX and PGP auxin transporter gene families in Sorghum bicolor under phytohormone and abiotic stress. The FEBS Journal. 2010;**277**(14):2954-2969

[29] Galvan-Ampudia CS, Julkowska MM, Darwish E, Gandullo J, Korver RA, Brunoud G, et al. Halotropism is a response of plant roots to avoid a saline environment. Current Biology. 2013;**23**(20):2044-2050. DOI: 10.1016/j.cub.2013.08.042

[30] Liu H, Liu B, Chen X, Zhu H, Zou C, Men S. AUX1 acts upstream of PIN2 in regulating root gravitropism. Biochemical and Biophysical Research Communications. 2018;**507**(1-4):433-436. DOI: 10.1016/j.bbrc.2018.11.056

[31] Swarup R, Friml J, Marchant A, Ljung K, Sandberg G, Palme K, et al. Localization of the auxin permease AUX1 suggests two functionally distinct hormone transport pathways operate in the Arabidopsis root apex. Genes & Development. 2001;**15**(20):2648-2653

[32] Van Den Berg T, Korver RA, Testerink C, Ten Tusscher KHWJ. Modeling Halotropism: A Key Role for Root Tip Architecture and Reflux Loop Remodeling in Redistributing Auxin. Development. 2016;**143**(18):3350-3362

[33] Michniewicz M, Zago MK, Abas L, Weijers D, Schweighofer A, Meskiene I, et al. Antagonistic Regulation of PIN Phosphorylation by PP2A and PINOID Directs Auxin Flux. Cell. 2007;**130**(6):1044-1056

[34] Grones P, Abas M, Hajný J, Jones A, Waidmann S, Kleine-Vehn J, et al. PID/WAG-mediated phosphorylation of the Arabidopsis PIN3 auxin transporter mediates polarity switches during gravitropism. Scientific Reports. 2018;**8**(1):1-11

[35] Sukumar P, Edwards KS, Rahman A, DeLong A, Muday GK. PINOID kinase regulates root gravitropism through modulation of PIN2-dependent basipetal auxin transport in arabidopsis. Plant Physiology. 2009;**150**(2):722-735

[36] Zhao Y. Auxin biosynthesis and its role in plant development. Annual Review of Plant Biology. 2010;**61**:49-64

[37] Ljung K, Hull AK, Celenza J, Yamada M, Estelle M, Normanly J, et al. Sites and regulation of auxin biosynthesis in arabidopsis roots. The Plant Cell. 2005;**17**(4):1090-1104

[38] Ryu H, Cho YG. Plant hormones in salt stress tolerance. Journal of Plant Biology. 2015;**58**(3):147-155

[39] Korver RA, Koevoets IT, Testerink C. Out of shape during stress: A key role for auxin. In: Trends in Plant Science. Vol. 23. Elsevier Ltd; 2018. pp. 783-793

[40] Smolko A, Bauer N, Pavlovi'c IP, Pěnčík A, Novák O, Salopek-Sondi B. Altered root growth, Auxin metabolism and distribution in Arabidopsis thaliana exposed to salt and osmotic Stress. International Journal of Molecular Sciences. 2021;**22**(15):7993. DOI: 10.3390/ijms22157993

[41] Tognetti VB, van Aken O, Morreel K, Vandenbroucke K, van de Cotte B, de Clercq I, et al. Perturbation of indole-3-butyric acid homeostasis by the UDP-glucosyltransferase UGT74E2 modulates Arabidopsis architecture and water stress tolerance. The Plant Cell. 2010;**22**(8):2660-2679

[42] Bajguz A, Piotrowska A. Conjugates of auxin and cytokinin. Phytochemistry. 2009;**70**(8):957-969. DOI: 10.1016/j.phytochem.2009.05.006

[43] Park JE, Park JY, Kim YS, Staswick PE, Jeon J, Yun J, et al. GH3-mediated auxin homeostasis links growth regulation with stress adaptation response in Arabidopsis. The Journal of Biological Chemistry. 2007;**282**(13):10036-10046. DOI: 10.1074/jbc.M610524200

[44] Xia XJ, Huang LF, Zhou YH, Mao WH, Shi K, Wu JX, et al. Brassinosteroids promote photosynthesis and growth by enhancing activation of rubisco and expression of photosynthetic genes in Cucumis sativus. Planta. 2009;**230**(6):1185-1196

[45] Li QF, He JX. BZR1 interacts with HY5 to mediate Brassinosteroid- and light-regulated cotyledon opening in Arabidopsis in darkness. Molecular Plant. 2016;**9**(1):113-125. DOI: 10.1016/j.molp.2015.08.014

[46] Gruszka D. The brassinosteroid signaling pathway-new key players and interconnections with other signaling networks crucial for plant development and stress tolerance. International Journal of Molecular Sciences. 2013;**14**(5):8740-8774

[47] Vert G, Nemhauser JL, Geldner N, Hong F, Chory J. Molecular mechanisms of steroid hormone signaling in plants. Annual Review of Cell and Developmental Biology. 2005;**21**:177-201

[48] Planas-Riverola A, Gupta A, Betegoń-Putze I, Bosch N, Ibanẹs M, Cano-Delgado AI. Brassinosteroid Signaling in Plant Development and Adaptation to Stress. Vol. 146. Development (Cambridge). Company of Biologists Ltd; 2019

[49] Gruszka D. Crosstalk of the brassinosteroid signalosome with phytohormonal and stress signaling components maintains a balance between the processes of growth and stress tolerance. International Journal of Molecular Sciences. 2018;**19**(9):2675-2719

[50] Gruszka D, Janeczko A, Dziurka M, Pociecha E, Oklestkova J, Szarejko I. Barley brassinosteroid mutants provide an insight into phytohormonal homeostasis in plant reaction to drought stress. Frontiers in Plant Science. 2016;7(December 2016):1-14

[51] Shiu SH, Karlowski WM, Pan R, Tzeng YH, Mayer KFX, Li WH. Comparative analysis of the receptor-like kinase family in Arabidopsis and rice. The Plant Cell. 2004;**16**(5):1220-1234

[52] Liu J, Zhang W, Long S, Zhao C. Maintenance of cell wall integrity under high salinity. International Journal of Molecular Sciences. 2021;**22**(6):1-19

[53] He J-X, Gendron JM, Yu S, Gampala SSL, Gendron N, Sun CQ, et al. BZR1 is a transcriptional repressor with dual roles in brassinosteroid homeostasis and growth responses. Science (80). 2005;**307**(March):1634-1638

[54] Sun X, Shantharaj D, Kang X, Ni M. Transcriptional and hormonal signaling control of Arabidopsis seed development. Current Opinion in Plant Biology. 2010;**13**(5):611-620. DOI: 10.1016/j.pbi.2010.08.009

[55] Yin Y, Vafeados D, Tao Y, Yoshida S, Asami T, Chory J. A new class of transcription factors mediates brassinosteroid-regulated gene expression in Arabidopsis. Cell. 2005;**120**(2):249-259

[56] Yu X, Li L, Zola J, Aluru M, Ye H, Foudree A, et al. A brassinosteroid transcriptional network revealed by genome-wide identification of BESI target genes in Arabidopsis thaliana. The Plant Journal. 2011;**65**(4):634-646

[57] Kumar V, Wani SH, Penna S, Tran LSP. Salinity responses and tolerance in plants, volume 1: Targeting sensory, transport and signaling mechanisms. Salin responses Toler Plants, vol 1 target sensory. Transpotation Signal Mechanical. 2018;**1**:1-399

[58] Ouertani RN, Arasappan D, Abid G, Ben CM, Jardak R, Mahmoudi H, et al. Transcriptomic analysis of salt-stress-responsive genes in barley roots and leaves. International Journal of Molecular Sciences. 2021;**22**(15):8155-8172

[59] Kagale S, Divi UK, Krochko JE, Keller WA, Krishna P. Brassinosteroid confers tolerance in Arabidopsis thaliana and Brassica napus to a range of abiotic stresses. Planta. 2007;**225**(2):353-364

[60] Anuradha S, Rao SR, S. Effect of brassinosteroids on salinity stress induced inhibition of seed germination and seedling growth of rice (Oryza sativa L.). Plant Growth Regulation. 2001;**33**(2):151-153

[61] Arora N, Bhardwaj R, Sharma P, Arora HK. 28-Homobrassinolide alleviates oxidative stress in salt-treated maize (Zea mays L.) plants. Brazilian. Journal of Plant Physiology. 2008;**20**(2):153-157

[62] Shahbaz M, Ashraf M. Influence of exogenous application of brassinosteroid on growth and mineral nutrients of wheat (Triticum aestivum l.) under saline conditions. Pakistan Journal of Botany. 2007;**39**(2):513-522

[63] Zeng H, Tang Q, Hua X. Arabidopsis Brassinosteroid mutants det2-1 and bin2-1 display altered salt tolerance. Journal of Plant Growth Regulation. 2010;**29**(1):44-52

[64] Chaiwanon J, Wang ZY. Spatiotemporal brassinosteroid signaling and antagonism with auxin pattern stem cell dynamics in Arabidopsis roots. Current Biology. 2015;**25**(8):1031-1042

[65] Tian H, Lv B, Ding T, Bai M, Ding Z. Auxin-BR interaction regulates plant growth and development. Frontiers in Plant Science. 2018;**8**:2256

[66] Piya S, Shrestha SK, Binder B, Neal Stewart C, Hewezi T. Protein-protein interaction and gene co-expression maps of ARFs and aux/IAAs in Arabidopsis. Frontiers in Plant Science. 2014;**5**(DEC):1-9

[67] Favero DS, Le KN, Neff MM. Brassinosteroid signaling converges with SUPPRESSOR OF PHYTOCHROME B4-#3 to influence the expression of SMALL AUXIN UP RNA genes and hypocotyl growth. The Plant Journal. 2017;**89**(6):1133-1145

[68] Östin A, Ilić N, Cohen JD. An in vitro system from maize seedlings for tryptophan-independent indole-3-acetic acid biosynthesis. Plant Physiology. 1999;**119**(1):173-178

[69] Zhang Y, Liang Z, Zong Y, Wang Y, Liu J, Chen K, et al. Efficient and transgene-free genome editing in wheat through transient expression of CRISPR/Cas9 DNA or RNA. Nature Communications. 2016;**7**:1-8

[70] Vriet C, Russinova E, Reuzeaua C. Boosting crop yields with plant steroids. The Plant Cell. 2012;**24**(3):842-857

[71] Nolan TM, Vukasinović N, Liu D, Russinova E, Yin Y. Brassinosteroids: Multidimensional regulators of plant growth, development, and stress responses. In: Plant Cell. American Society of Plant Biologists; 2020. pp. 298-318. DOI: 10.1105/tpc.19.00335

[72] Wu G, Wang X, Li X, Kamiya Y, Otegui MS, Chory J. Methylation of a phosphatase specifies dephosphorylation and degradation of activated brassinosteroid receptors. Science Signaling. 2011;**4**(172):172-197

[73] Bradai M, Amorim-Silva V, Belgaroui N, Esteban Del Valle A, Chabouté M-E, Schmit A-C, et al. Wheat type one protein phosphatase participates in the brassinosteroid control of root growth via activation of BES1. International Journal of Molecular Sciences. 2021;**19**:10424. DOI: 10.3390/ijms221910424

[74] Hacham Y, Sela A, Friedlander L, Savaldi-Goldstein S. BRI1 activity in the root meristem involves post-transcriptional regulation of PIN auxin efflux carriers. Plant Signaling & Behavior. 2012;**7**(1):5-8

[75] Chung Y, Maharjan PM, Lee O, Fujioka S, Jang S, Kim B, et al. Auxin stimulates DWARF4 expression and brassinosteroid biosynthesis in Arabidopsis. The Plant Journal. 2011;**66**(4):564-578

[76] Saini S, Sharma I, Kaur N, Pati PK. Auxin: A master regulator in plant

root development. Plant Cell Reports. 2013;**32**(6):741-757

[77] Salazar-Henao JE, Lehner R, Betegón-Putze I, Vilarrasa-Blasi J, Caño-Delgado AI. BES1 regulates the localization of the brassinosteroid receptor BRL3 within the provascular tissue of the Arabidopsis primary root. Journal of Experimental Botany. 2016;**67**(17):4951-4961. Available from: https://academic.oup.com/jxb/article/67/17/4951/2197600

[78] Souter MA, Pullen ML, Topping JF, Zhang X, Lindsey K. Rescue of defective auxin-mediated gene expression and root meristem function by inhibition of ethylene signalling in sterol biosynthesis mutants of Arabidopsis. Planta. 2004;**219**(5):773-783

[79] Nakamura A. Brassinolide induces IAA5, IAA19, and DR5, a synthetic auxin response element in Arabidopsis, implying a cross talk point of Brassinosteroid and auxin Signaling. Plant Physiology. 2003;**133**(4):1843-1853

[80] Vert G, Walcher CL, Chory J, Nemhauser JL. Integration of auxin and brassinosteroid pathways by auxin response factor 2. Proceedings of the National Academy of Sciences of the United States of America. 2008;**105**(28):9829-9834

[81] Cho H, Ryu H, Rho S, Hill K, Smith S, Audenaert D, et al. A secreted peptide acts on BIN2-mediated phosphorylation of ARFs to potentiate auxin response during lateral root development. Nature Cell Biology. 2014;**16**(1):66-76

[82] Zhou Y, Zhang X, Kang X, Zhao X, Zhang X, Ni M. Short hypocotyl under Blue1 associates with Miniseed3 and Haiku2 promoters in vivo to regulate arabidopsis seed development. The Plant Cell. 2009;**21**(1):106-117

[83] Aki OM, Floyd SK, Fischer RL, Goldberg RB, Harada JJ. Effects of APETALA2 on embryo, endosperm, and seed coat development determine seed size in Arabidopsis. Sexual Plant Reproduction. 2009;**22**(4):277-289

[84] Kim H, Park PJ, Hwang HJ, Lee SY, Oh MH, Kim SG. Brassinosteroid signals control expression of the AXR3/IAA17 gene in the cross-talk point with auxin in root development. Bioscience, Biotechnology, and Biochemistry. 2006;**70**(4):768-773

[85] Yan L, Ma Y, Liu D, Wei X, Sun Y, Chen X, et al. Structural basis for the impact of phosphorylation on the activation of plant receptor-like kinase BAK1. Cell Research. 2012;**22**(8):1304-1308

[86] Mano Y, Nemoto K. The pathway of auxin biosynthesis in plants. Journal of Experimental Botany. 2012;**63**(8):2853-2872

[87] Stepanova AN, Robertson-Hoyt J, Yun J, Benavente LM, Xie DY, Doležal K, et al. TAA1-mediated auxin biosynthesis is essential for hormone crosstalk and plant development. Cell. 2008;**133**(1):177-191

[88] Tao Y, Ferrer JL, Ljung K, Pojer F, Hong F, Long JA, et al. Rapid synthesis of auxin via a new tryptophan-dependent pathway is required for shade avoidance in Plants. Cell. 2008;**133**(1):164-176

[89] Cackett L, Cannistraci CV, Meier S, Ferrandi P, Pěnčík A, Gehring C, et al. Salt-specific gene expression reveals elevated auxin levels in Arabidopsis thaliana Plants grown under saline conditions. Frontiers in Plant Science. 2022;**13**:804716

[90] Ren H, Gray WM. SAUR proteins as effectors of hormonal and environmental signals in plant Growth. Molecular Plant. 2015;**8**(8):1153-1164. DOI: 10.1016/j.molp.2015.05.003

[91] Leyser HMO, Lincoln CA, Timpte C, Lammer D, Turner J, Estelle M. Arabidopsis auxin-resistance gene AXR1 encodes a protein related to ubiquitin-activating enzyme E1. Nature. 1993;**364**(6433):161-164

[92] Kirungu JN, Magwanga RO, Lu P, Cai X, Zhou Z, Wang X, et al. Functional characterization of GH-A08G1120 (GH3.5) gene reveal their significant role in enhancing drought and salt stress tolerance in cotton. BMC Genetics. 2019;**20**(1):1-17

[93] Omelyanchuk NA, Kovrizhnykh VV, Oshchepkova EA, Pasternak T, Palme K, Mironova VV. A detailed expression map of the PIN1 auxin transporter in Arabidopsis thaliana root. BMC Plant Biology. 2016;**16**(1):1-12. DOI: 10.1186/s12870-015-0685-0

[94] Yuan X, Xu P, Yu Y, Xiong Y. Glucose-TOR signaling regulates PIN2 stability to orchestrate auxin gradient and cell expansion in Arabidopsis root. Proceedings of the National Academy of Sciences of the United States of America. 2020;**117**(51):32223-32225

[95] Péret B, Middleton AM, French AP, Larrieu A, Bishopp A, Njo M, et al. Sequential induction of auxin efflux and influx carriers regulates lateral root emergence. Molecular Systems Biology. 2013;**9**(699):1-15

[96] Friml J, Benková E, Blilou I, Wisniewska J, Hamann T, Ljung K, et al. AtPIN4 mediates sink-driven auxin gradients and root patterning in Arabidopsis. Cell. 2002;**108**(5):661-673

[97] Mravec J, Skůpa P, Bailly A, Hoyerová K, Křeček P, Bielach A, et al. Subcellular homeostasis of phytohormone auxin is mediated by the ER-localized PIN5 transporter. Nature. 2009;**459**(7250):1136-1140

[98] Cazzonelli CI, Vanstraelen M, Simon S, Yin K, Carron-Arthur A, Nisar N, et al. Role of the Arabidopsis PIN6 auxin transporter in auxin homeostasis and auxin-mediated development. PLoS One. 2013;**8**(7)

[99] Sun N, Wang J, Gao Z, Dong J, He H, Terzaghi W, et al. Arabidopsis saurs are critical for differential light regulation of the development of various organs. Proceedings of the National Academy of Sciences of the United States of America. 2016;**113**(21):6071-6076

[100] Spartz AK, Lee SH, Wenger JP, Gonzalez N, Itoh H, Inzé D, et al. The SAUR19 subfamily of SMALL AUXIN UP RNA genes promote cell expansion. The Plant Journal. 2012;**70**(6):978-990

[101] Zhang J, Peer WA. Auxin homeostasis: The DAO of catabolism. Journal of Experimental Botany. 2017;**68**(12):3145-3154

[102] Okushima Y, Fukaki H, Onoda M, Theologis A, Tasaka M. ARF7 and ARF19 regulate lateral root formation via direct activation of LBD/ASL genes in Arabidopsis. The Plant Cell. 2007;**19**(1):118-130

[103] Lim PO, Lee IC, Kim J, Kim HJ, Ryu JS, Woo HR, et al. Auxin response factor 2 (ARF2) plays a major role in regulating auxin-mediated leaf longevity. Journal of Experimental Botany. 2010;**61**(5):1419-1430

[104] Wójcikowska B, Gaj MD. Expression profiling of AUXIN RESPONSE FACTOR genes during somatic embryogenesis induction in Arabidopsis. Plant Cell Reports. 2017;**36**(6):843-858

[105] Choi HS, Seo M, Cho HT. Two TPL-binding motifs of ARF2 are involved in repression of auxin responses. Frontiers in Plant Science. 2018;**9**(March):1-9

[106] Mai YX, Wang L, Yang HQ. A gain-of-function mutation in IAA7/AXR2 confers late flowering under Short-day light in Arabidopsis. Journal of Integrative Plant Biology. 2011;**53**(6):480-492

[107] Wei T, Zhang L, Zhu R, Jiang X, Yue C, Su Y, et al. A gain-of-function mutant of iaa7 inhibits stem elongation by transcriptional repression of expa5 genes in brassica napus. International Journal of Molecular Sciences. 2021;**22**(16)

[108] Fukaki H, Tameda S, Masuda H, Tasaka M. Lateral root formation is blocked by a gain-of-function mutation in the SOLITARY-ROOT/IAA14 gene of Arabidopsis. The Plant Journal. 2002;**29**(2):153-168. DOI: 10.1046/j.0960-7412.2001.01201.x

[109] Aslam M, Sugita K, Qin Y, Rahman A. Aux/iaa14 regulates microrna-mediated cold stress response in arabidopsis roots. International Journal of Molecular Sciences. 2020;**21**(22):8441

[110] da Costa CT, Offringa R, Fett-Neto AG. The role of auxin transporters and receptors in adventitious rooting of Arabidopsis thaliana pre-etiolated flooded seedlings. Plant Science. 2020:290

[111] Swarup R, Bhosale R. Developmental roles of AUX1/LAX auxin influx carriers in Plants. Frontiers in Plant Science. 2019;**10**(October):1-14

[112] Sun L, Feraru E, Feraru MI, Waidmann S, Wang W, Passaia G, et al. PIN-LIKES coordinate brassinosteroid signaling with nuclear auxin input in Arabidopsis thaliana. Current Biology. 2020;**30**(9):1579-1588

[113] Choe S, Dilkes BP, Fujioka S, Takatsuto S, Sakurai A, Feldmann KA. The DWF4 gene of Arabidopsis encodes a cytochrome P450 that mediates multiple 22α-hydroxylation steps in brassinosteroid biosynthesis. The Plant Cell. 1998;**10**(2):231-243

[114] Tanaka K, Asami T, Yoshida S, Nakamura Y, Matsuo T, Okamoto S. Brassinosteroid homeostasis in Arabidopsis is ensured by feedback expressions of multiple genes involved in its metabolism. Plant Physiology. 2005;**138**(2):1117-1125

[115] Li QF, Lu J, Yu JW, Zhang CQ, He JX, Liu QQ. The brassinosteroid-regulated transcription factors BZR1/BES1 function as a coordinator in multisignal-regulated plant growth. In: Biochimica et Biophysica Acta - Gene Regulatory Mechanisms. Vol. 1861. no. 6. Elsevier B.V; 2018. pp. 561-571

[116] Ren H, Wu X, Zhao W, Wang Y, Sun D, Gao K, et al. Heat shock-induced accumulation of the glycogen synthase kinase 3-like kinase BRASSINOSTEROID INSENSITIVE 2 promotes early flowering but reduces thermotolerance in Arabidopsis. Frontiers in Plant Science. 2022;**13**(January):1-14

[117] Wang R, Wang J, Zhao L, Yang S, Song Y. Impact of heavy metal stresses on the growth and auxin homeostasis of Arabidopsis seedlings. Biometals. 2015;**28**(1):123-132

[118] Cackett L, Luginbuehl LH, Schreier TB, Lopez-Juez E, Hibberd JM. Chloroplast development in green plant tissues: the interplay between light, hormone, and transcriptional regulation. The New Phytologist. 2022;**233**(5):2000-2016

[119] Oh E, Zhu JY, Bai MY, Arenhart RA, Sun Y, Wang ZY. Cell elongation is regulated through a central circuit of interacting transcription factors in the Arabidopsis hypocotyl. eLife. 2014;**2014**(3):1-19

[120] Khan Y, Xiong Z, Zhang H, Liu S, Yaseen T, Hui T. Expression and roles of GRAS gene family in plant growth, signal transduction, biotic and abiotic stress resistance and symbiosis formation - A review. Plant Biology. 2022;**24**(3):404-416

[121] Cabello JV, Lodeyro AF, Zurbriggen MD. Novel perspectives for the engineering of abiotic stress tolerance in plants. Current Opinion in Biotechnology. 2014;**26**:62-70. DOI: 10.1016/j.copbio.2013.09.011

[122] Iqbal S, Wang X, Mubeen I, Kamran M, Kanwal I, Díaz GA, et al. Phytohormones trigger drought tolerance in crop Plants: Outlook and future perspectives. Frontiers in Plant Science. 2022;**12**(January):1-14

[123] Feng Y, Yin Y, Fei S. Down-regulation of BdBRI1, a putative brassinosteroid receptor gene produces a dwarf phenotype with enhanced drought tolerance in Brachypodium distachyon. Plant Science. 2015;**234**:163-173. DOI: 10.1016/j.plantsci.2015.02.015

[124] Hayat S, Ahmad A. Brassinosteroids: A Class of Plant Hormone. Brassinosteroids: A Class of Plant Hormone; 2011. pp. 1-462

[125] Hwang H, Ryu H, Cho H. Brassinosteroid signaling pathways interplaying with diverse signaling cues for crop enhancement. Agronomy. 2021;**11**(3):1-12

[126] Yu X, Han J, Wang E, Xiao J, Hu R, Yang G, et al. Genome-wide identification and homoeologous expression analysis of PP2C genes in wheat (Triticum aestivum L.). Frontiers in Genetics. 2019;**10**(561)

[127] Peng Z, Wang Y, Geng G, Yang R, Yang Z, Yang C, et al. Comparative analysis of Physiological, enzymatic, and transcriptomic responses revealed mechanisms of salt tolerance and recovery in Tritipyrum. Front. Plant Science. 2022:12

[128] Martins S, Montiel-Jorda A, Cayrel A, Huguet S, Paysant-Le Roux C, Ljung K, et al. Brassinosteroid signaling-dependent root responses to prolonged elevated ambient temperature. Nature Communications. 2020;**8**(1):309. Available from: http://public.wmo.int

[129] Liu JM, Park SJ, Huang J, Lee EJ, Xuan YH, Il JB, et al. Loose plant Architecture1 (LPA1) determines lamina joint bending by suppressing auxin signalling that interacts with C-22-hydroxylated and 6-deoxo brassinosteroids in rice. Journal of Experimental Botany. 2016;**67**(6):1883-1895

[130] Gruszka D. Exploring the brassinosteroid signaling in monocots reveals novel components of the pathway and implications for plant breeding. International Journal of Molecular Sciences. 2020;**21**(1):354

Chapter 7

Achieving Salinity-Tolerance in Cereal Crops: Major Insights into Genomics-Assisted Breeding (GAB)

Ram Baran Singh and Rajni Devi

Abstract

Cereal crops including rice, wheat, corn, sorghum, pearl millet and small millet, are grown for food, feed and fuel in crop-livestock based agricultural systems around the world. Soil salinity occupies an important place among the soil problems that threaten the sustainability of agriculture in a wide area around the world. Salinity intensity is predicted to exacerbate further due to global warming and climate change, requiring greater attention to crop breeding to increase resilience to salinity-induced oxidative stress. Knowledge of physiological responses to varying degrees of oxidative stress has helped predict crop agronomic traits under saline ecosystems and their use in crop breeding programs. Recent developments in high-throughput phenotyping technologies have made it possible and accelerated the screening of vast crop genetic resources for traits that promote salinity tolerance. Many stress-tolerant plant genetic resources have been developed using conventional crop breeding, further simplified by modern molecular approaches. Considerable efforts have been made to develop genomic resources which used to examine genetic diversity, linkage mapping (QTLs), marker-trait association (MTA), and genomic selection (GS) in crop species. Currently, high-throughput genotyping (HTPG) platforms are available at an economical cost, offering tremendous opportunities to introduce marker-assisted selection (MAS) in traditional crop breeding programs targeting salinity. Next generation sequencing (NGS) technology, microenvironment modeling and a whole-genome sequence database have contributed to a better understanding of germplasm resources, plant genomes, gene networks and metabolic pathways, and developing genome-wide SNP markers. The use of developed genetic and genomic resources in plant breeding has paved a way to develop high yielding, nutrient-rich and abiotic stress tolerant crops. Present chapter provides an overview of how the strategic usage of genetic resources, genomic tools, stress biology, and breeding approaches can further enhance the breeding potential and producing salinity-tolerant crop varieties/lines.

Keywords: environmental stress, stress tolerance mechanisms, traditional and modern breeding, plant genetic and genomic resources marker-assisted selection (MAS), genomic prediction (GP)

1. Introduction

Food is one of the basic needs of every organism including human being to survive hence its importance cannot be exaggerated. Beyond the need to quench appetite, food is extremely essential for optimal functioning of the entire body physiology and metabolism. All requirements meet out by a balanced diet that comprise essential macro and micro nutrients available in several types of food ingredients (i.e., carbohydrates, proteins, lipids, mineral and vitamins, and dietary fibers) consumed as daily meal [1]. Among all the necessary food ingredients, carbohydrates are the key ingredients occurring in balanced diets as a primary source of energy required to perform routine workout and other physical actions. Most of the carbohydrates are supplied by cereal grains of the grassy crops [2]. Cereals which include; rice, wheat, corn, sorghum, pearl millet and small millets have been grown since time immemorial for food, fodder and fuel in crop-livestock based agricultural systems around the world [3]. Most of the cereals have predominantly been considered as a staple foods in every agro-ecology and promoted as a healthy food in body weight management, however each cereal contribute in different ways. Plant based foods in various forms are an essential components in human diets which contain essential ingredients. For instance, millets are high-quality alternative to major cereals (rice, wheat, and maize) owing to their greater minerals and proteins contents. In addition, statistical studies have shown that half of the total percentage of calories consumed by human population comes from cereals which are the most traded agricultural commodity in international markets [4].

Global cereal growing area is projected to increase by 14 million hectares between 2020 and 2030 and harvested area over developed countries projected to grow by 4 million hectares, in Russia, Ukraine and Australia, as well as in developing countries using about 10 million hectares, mainly in Asian and Latin American countries. Arable land area under wheat and corn is expected to be raised by ~3% and 4%, however other areas under coarse grains and rice expected to remain unchanged. As land expansion is constrained by limited arable land accessibility compared to the previous decade, resulting restrictions placed on the conversion of forests or pastures to cropland, and continued urbanization, worldwide production growth is predictable to be driven mostly by intensification. Yield growth due to improved cultivation technology and methods, particularly in developing countries, is predicted to support prospect cereal production. Thus, the upward trend in cereal production observed in recent decades is indicative of the progress made in the agricultural sector around the world (**Figure 1**). Global yields are expected to increase between the base period and 2030 by approx 9% for wheat and other coarse grains, 10% for corn and 12% for rice. World wheat production is anticipated to hike by 87 million tons to 840 million tons by 2030, a moderate pace in relative terms compared to the past decade. India is the world's third-largest wheat producer, expected to provide the largest share of additional wheat supplies, boosting production by 18 million tonnes by 2030 and expanding acreage in response to a national policy to increase self-sufficiency in wheat. Nevertheless, consumption of the cereal grains is greater than the production; hence a higher cereal production with improved nutritional values will need to meet out increasing demands due the burgeoning population. Global cereal production predicted to be lesser than consumption requirements in future due to several environmental stresses in the wake of climate change that may lead to drawdown in cereal stock globally.

Among all the cereal, rice (*Oryza sativa*) is major staple food grain belongs to family *Poaceae* worldwide and has extensive economic importance. Rice provides feed

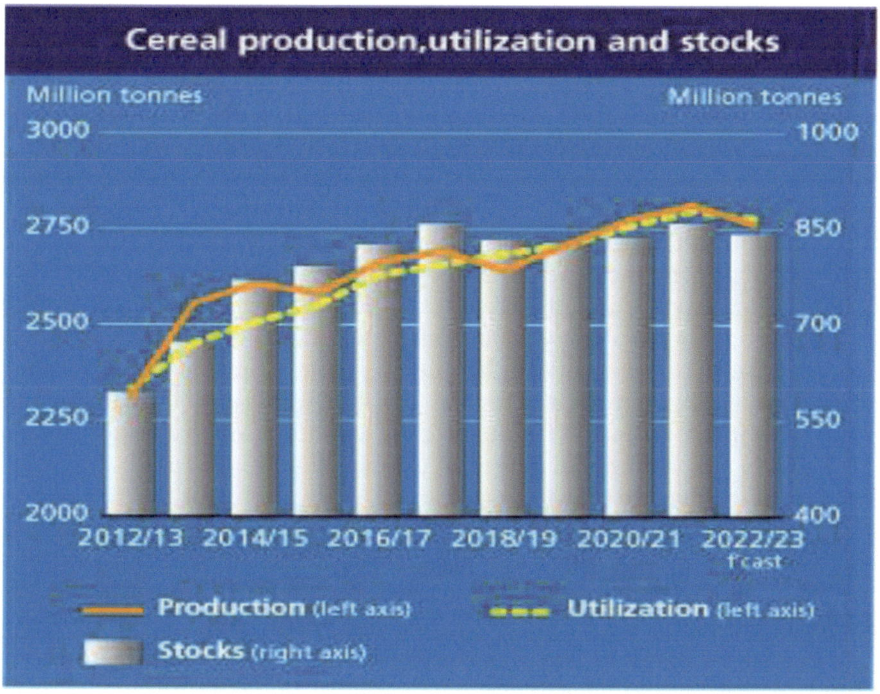

Figure 1.
Bar diagram illustrating the year wise production and consumption of cereal grains globally.

to more than 50% of world's population predominantly in Asia where the population is expected to rise from 4.3 to 5.2 billion by 2050. Wheat (*Triticum aestivum L.)* is the second most important staple cereal crop grown all over the world contributing substantially to the world's food and nutritional security world [5]. About 20–30% daily calorie intake [6] and 55% of carbohydrates are provided by wheat worldwide [7]. Wheat bread has high vitamins B, thiamine, and B2-riboflavin content with other several minor nutrients [8]. Worldwide estimated annual production of wheat is about 768.90 million metric tons whereas; India's contribution is about 107.6 million metric tons [9]. Maize (*Zea mays*) is a cereal grain belongs to family Poaceae, cultivated throughout the world. The maize production globally in the year 2020–2021 was 1.2 billion tonnes led by India with 30.2 million metric tonnes [9]. Maize is a staple food plays an important role in food and nutritional security worldwide. Sorghum (*Sorghum bicolar*) is a cereal crop used for grain, fiber and fodder. Sorghum is cultivated in warmer climates worldwide and nutritional profile includes several minerals like phosphorous, iron zinc and copper it is also a good source of b-complex and vitamins. It estimated that the world sorghum production in year 2021–2022 was 60.32 million tons whereas; India's contribution in total production was 4.7 million tons. Pearl millet (*Pennisetum glaucum L.*) has been widely grown in Sub-Saharan Africa, South Asia and Indian subcontinent. Archeological proof and modern genome sequence analysis showed that pearl millet originated and domesticated about 4000 to 5000 years ago in West Africa [10]. The global millet production in the year 2020–2021 was 30.5 million metric tonnes led by India with 41% (12.5) million metric tons [11]. Pearl millet is one the most important eco-friendly field crop in conventional farming system, plays an important role in food and nutritional security [3].

However, multiple types of environmental stresses (abiotic and biotic) are affecting holistic growth and plant health which ultimately leads to low crop productivity globally. Among the abiotic stresses, soil salinity or salt stress is one the most brutal environmental stresses adversely affecting sustainable crop yield and productivity globally. Hence, a comprehensive research on abiotic stresses in cereals has been experienced to cope with such a crucial salinity problem and related breakthroughs have been conducted and proved to be as one of the key activities to obtain higher genetic gains in terms of adaptability in changing climatic scenarios. Salinity adversely impacts on plant growth and development by exerting ion cytotoxicity, osmotic shock, nutrient imbalance, and oxidative stress [12]. The influence of soil salinity leads to impaired plant physiology, biochemistry and metabolism, hormonal imbalance, and regulatory pathways at cellular or entire plant structure. Plants response to the access salt stress comprised of several physiological and molecular approaches operates in cells in a coordinated fashion to cope with ion toxicity and hyperosmolarity [13]. A plenty of conventional as well as advanced molecular approaches have been developed and employed to breed the salinity tolerant varieties of the different cereal crops [14].

2. Abiotic stresses

Multiple abiotic (including; drought, heat, salinity, flood, harmful radiation, heavy metals, gaseous pollutants) and biotic stresses that are attacks of different microbial pathogens (e.g., fungi, bacteria, viruses, viroids, oomycetes, nematodes, and phtyotoplasma) causing adverse effects on morphology, physiology and metabolism leads to impaired plant growth and yield potential. Abiotic stress is progressively predominant in the wake of climate change and global warming affecting overall plants growth at different developmental stages such as germination, vegetative, and reproductive phase [15]. Total yield of crop is immensely affected by various factors like climatic fluctuations, insect incidents, agronomic factors, and nutrient availability in the soil. Stress is the any adverse environmental condition that hampers optimal growth and development of the plants [16]. Crop productivity and adaptability is affected by mainly physiological heat, drought, salinity and cold oxidative stresses. According to statistical estimates, approximately 20% of agricultural land is under salt stress, which negatively affects plant physiology and, ultimately, yield and nutritional value. The role of identified germplasm has been emphasized for drought breeding as the measured performance under drought stress is largely a result of adaptation to stress conditions. Hybridization of adapted landraces with selected elite genetic material has been testified to amalgamate adaptation and productivity. Abiotic stress is governed by quantitative trait hence genes linked to these traits have been identified and used to select desirable alleles responsible for stress tolerance in plant. Abiotic stress reduce water availability to plant roots by increasing water soluble salts in soil and plants suffer from increased osmotic pressure outside the root. Physiological changes include lowering of leaf osmotic potential, water potential and relative water content, creation of nutritional imbalance, enhancing relative stress injury or one or more combination of these factors. Plants operate a number of molecular, cellular and physiological modifications to overcome abiotic stresses [13]. Morphological and biochemical changes include changes in root and shoot length, number of leaves, secondary metabolite (glycine betaine, proline, malondialdehyde (MDA), abscisic acid) accumulation in plant, source and sink ratio.

Plants have developed various mechanisms in order to overcome these threats of biotic and abiotic stresses. They sense the external stress environment, get stimulated and then generate appropriate cellular responses. They perform this by stimuli received from the sensors located on the cell surface or cytoplasm and transmitted to transcriptional machinery situated in nucleus through various signal transduction pathways. This leads to differential transcriptional changes making plant tolerant against stress. Signaling pathways act as a connecting link and play an important role between sensing stress environment and generating an appropriate biochemical and physiological response.

2.1 Salinity stress

Soil salinity or salt stress refers to concentrations of salts in soils that affect physiological, biochemical process, growth and productivity of crops [17]. Salinity is one of the most important challenges which induce various physiological, molecular and cellular responses in plants [12]. Salt stress leads to loss of production by aggregation of soluble salts in the soil of root zone (18) which finally cause about 10 mha annual losses of arable land. A large number of sodium ions (Na^+), carbonate and bicarbonate anions present in exchange sites affecting pH ranges [18]. Approximately 900 million hectares of agricultural land are affected by salinity globally, with India contributing 7 million hectares [19]. Salt stress affects approx 32 million hectares out of 1500 million hectares of dry agriculture land and 45 million hectares of irrigated land [20]. Therefore, yield reduction because of the increase in salt stress will have a disproportionately large effect.

Based on the salinity stress, three types of the soils are found in different geographical areas [21], which includes; (i) Saline soils which contains high level of water-soluble salt and electro-conductivity (EC) exceeded up to 4ds/m. By definition, saline soil has EC $\geq$ of 4 dSm^{-1} (equal to approximately 40 mM NaCl), whereas soils are considered strongly saline if the EC $\geq$ of 15 dSm^{-1}. (ii) Sodic soils contain high contents of exchangeable sodium on the cation-exchange sites and usually have pH values ranges from 7.0 to 8.5. There are high accumulation of Na^+ which cause soil collides to disperse cations (Ca^{+2} and Mg^{+2}) insufficiently. Distributed collides clog the soil's pore and reduce the ability to transport water and air. Sodic soils have the tendency to extreme swelling and shrinking during moist and dry conditions respectively. (iii) Saline-sodic soil which is known for high producing soil at the arid and semi-arid area. This soil shows dual nature that associates with high EC (>4 dSm^{-1}) and low pH (below 8.5), hence both excess salts and Na^+ affects plant growth under in saline-sodic soil conditions. Under salinity, plant growth is affected by two-phase salt stress.

2.1.1 Phases of salt stress

2.1.1.1 Osmotic phase

In osmotic phase, the excess soil salt concentration reduced water potential in root zone causing water scarcity results in impaired plant growth. Growth reduction in osmotic phase rely only on the outer surface of salt concentration not in inside the plant tissues. A primary cause of reduced growth is that plant has to expense major portion of energy to acquire water from salty soil for routine metabolic process (**Figure 2**). Due to osmotic stress low water potential occurs in salty soil for plant

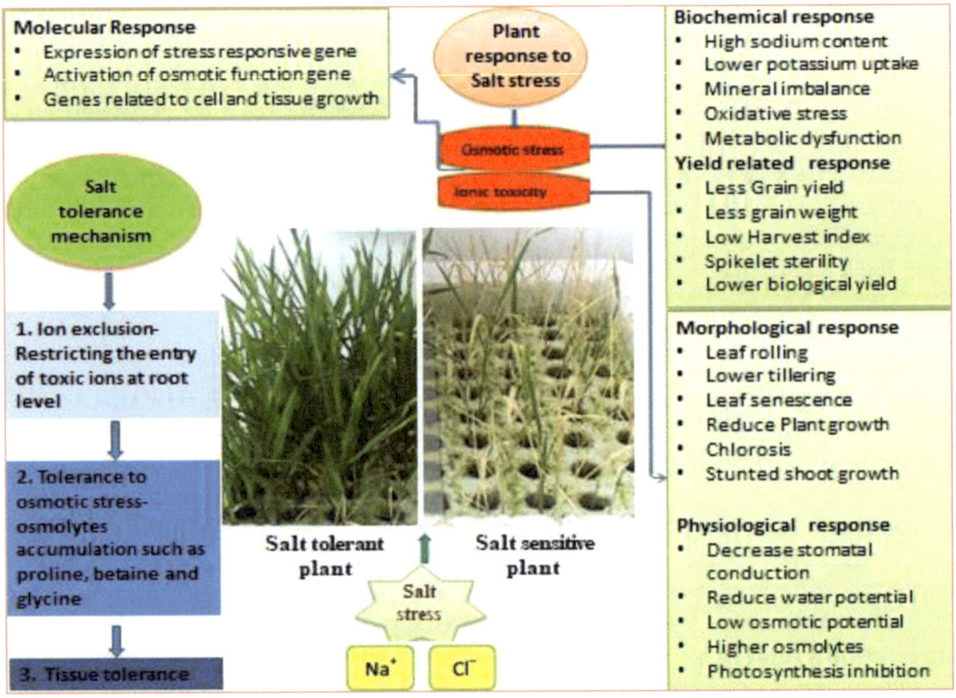

Figure 2.
Diagram displaying plant responses and mechanism of salt tolerance showing ion exclusion, osmotic tolerance and tissue tolerance in crops.

uptake even if the volumetric soil water content is higher as field capacity. Osmotic stress initially causes numerous physiological changes, like membranes disruption, nutrient imbalance, decreased photosynthetic activity, and reduction of the stomatal aperture [12].

2.1.1.2 Ion phase

Reduction of plant growth in the ionic phase is mainly due to internal tissue injury caused due to high accumulation of toxic Na^+. In addition, salt stress leads to reduced plant growth due to higher intake of certain ions (Na^+ and Cl^-) which is known as ion toxicity [22]. Accumulation of Na^+ and Cl^- in higher concentration delays the growth and negatively affects the metabolic processes in plants. The Na^+ ions inhibits K^+ ion uptake in plants and disturbs stomatal regulation which is the reason for water loss and necrosis. The Cl^- ions stimulates chloride toxicity due to defective production of chlorophyll.

2.1.2 Mechanism of salt tolerance in plants

Tolerance to salt stress varies on species and variety and it also varies at growth stages of plants. A wide spectrum of responses against salt stress shown by plant gives warranty of a broad range of adaptations at the entire plant level [23]. Many crops tolerate negative effect of salinity and could survive with their routine function and metabolisms. Salt tolerance is associated with accumulation of compatible osmolytes and exclusion of Na^+ while entering into the plant or tissue tolerance to high Na^+ [24].

To grow and reproduce in high salt stress conditions plants had developed several mechanisms (**Figure 2**) these mechanisms can be categorized into three steps [25]; (i) ions exclusion, it regulates Na^+ and Cl^- uptake. it prevent the accumulation of toxic ions into leaves (ii) osmotic stress tolerance, limits the growth of stems which is controlled by long-distance signals and is activated prior shoot Na^+ accumulation; and (iii) tissue tolerance, tolerance power of tissues against accumulated Na^+ or Cl^-. When Na^+ or Cl^- ions get succeeded to enter inside the plant tissue it grouped in leaf vacuole to prevent it from salt injury of thylakoid membrane.

2.1.2.1 Ion exclusion

Intercellular compartmentalization of Na^+ ion is a main tolerance mechanism, which provides the ability to leaves to bear with high Na^+ concentration (**Figure 3**). Salt stress imbalance the ion ratio by altering the pathway of sodium intake in place of potassium acquisition. There are four mechanisms for Na^+ exclusion [26]; (i) Selective permeability of ions in cortex and stele by root cells, (ii) Stocking of xylem in root by xylem parenchyma cells, (iii) Elimination of salt from the stem by xylem parenchyma cells and, (iv) phloem stacking. Salt entering through root system can be excluded, or inside the plant system salt can be barred from entering sensitive organs. Most of the plant species grown under salt stress, Na^+ ion competes with Cl^- to reach a concentration of toxic level. Therefore, the main focus of the researcher is on Na^+ exclusion and transport within plant. Na^+ exclusion by roots regulates toxicity level of Na^+ within leaf blade; however, fails to do that induce Na^+ toxicity after a short or long period, and cause death of older leaves. Exclusion of Na^+ at root level occurs as a result of ion selectivity. A mechanism that operates for uptake of Na^+ and K^+ ions reported earlier by Schachtman and Schroeder [27]. Salt tolerance in many species is associated with a high concentration of K^+ in young expanding leaf tissues. So it clearly shows the possibility of association of Na^+/K^+

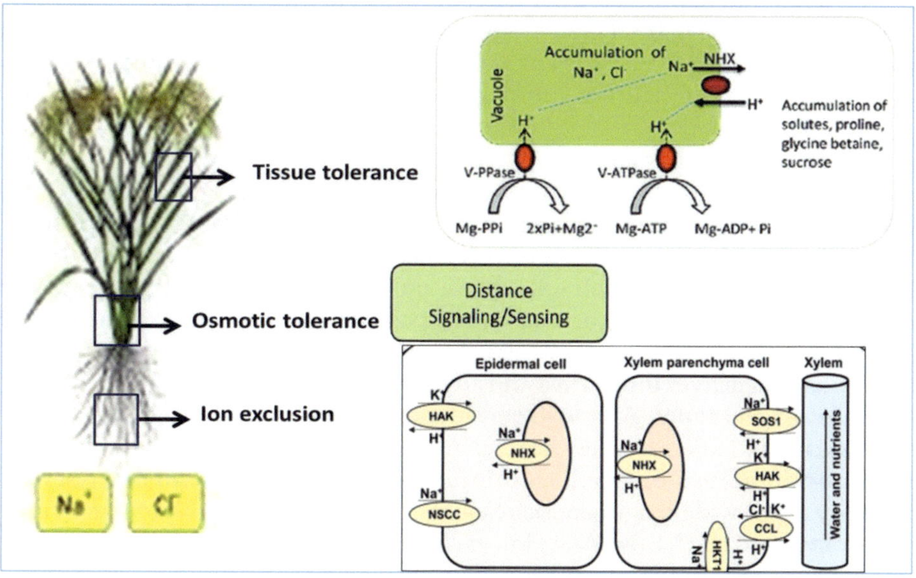

Figure 3.
Diagrammatic representation of all the abiotic stress tolerance mechanisms operated in plants.

with salt tolerance. The control of Na^+ uptake and better maintenance of the K+/ Na^+ ratio can be considered as key cellular mechanism to maintain osmotic potential for optimum cell activities, which contributes provides better adaptation capacity in plant under stress conditions [28, 29].

2.1.2.2 Tolerance to osmotic stress

Osmotic tolerance is reducing the osmotic potential due to solute accumulation in response to water stress, plays key role in plant adaptation to dehydration by maintaining turgor pressure, relative water content, and high stomatal conductance [30]. The osmotic effect is measured by growth rate and stomatal conductance of plants. Proline is a widely distributed osmolyte which protects the plant cells against salt stress. Proline acts as osmolyte that protects subcellular structure and biomolecules and chelate metal ions under osmotic stress [31]. In drought conditions, the delayed stomatal closure caused mainly by higher osmotic adjustment led to elevated assimilation rate and assimilates production [32]. In addition, increase in polyphenol content, glycine betaine and antioxidant enzymes activities have been reported associated with stress tolerance in plants providing protection against reactive oxygen species (ROS) [33]. All the possible mechanisms of salt tolerance operate at cellular levels in plants are depicted in **Figure 3**.

2.1.2.3 Tissue tolerance

Third mechanism, as tissue tolerance is increases the shelf-life of older leaves, where it distributes Na^+ and Cl^- at intercellular level to remove toxic levels of ions from cytoplasm and leaf mesophyll cells and to avoid detrimental effect on cellular process [25]. It synthesizes compatible solutes and controls transport and biochemical processes in cytoplasm and thus has dual function as osmoprotectant and osmotic adjustment (lowering of osmotic potential) [34]. These compatible solutes regulate osmotic tolerance in plants through different pathways such as it protects enzyme to get denature, stabilize plasma membrane and regulate macromolecules by osmotic adjustment [35].

3. Breeding approaches for salinity tolerance

Plants adapted specific mechanisms to tolerate saline conditions and activate various genes for salt tolerance to counter osmotic and oxidative stresses induced by salinity. Genetic evolution of salt tolerance is quite complex, while improvement has made less progress than anticipation over the past few decades. The explosive generation of information and technology related to genetics and genomics over the past decades pledge to deliver innovative and advanced resources for the potential production of tolerant genotypes. Although considerable progress in defining the primary mechanisms of salt tolerance, key hurdles are yet to be resolved in the translation and combination of the resulting molecular information into the plant breeding activities. Availability of the wide range genetic resources in cereals along with the implementation of advanced mechanisms like ion exclusion, osmatic tolerance, and tissue tolerance have been continuously improving screening procedures for salt tolerance genotypes. It could be enhanced via using traditional breeding or molecular breeding techniques such pyramiding, introgression by employing the genes or alleles which are already reported to have the potential of salt tolerance in the crop plants. Thus, considering these advantages, several breeders have been employed the techniques in cereal crop for salt tolerance enhancement.

3.1 Traditional cereal breeding for salinity tolerance

Plant breeding is decisive manipulation of plant species to produce desired plant accessions that are better suited for cultivation, produce higher yields, and stress resistance. Breeders selected edible plants with certain desirable traits and over time these valuable traits accumulated. Initial period of plant breeding spans from beginning of agriculture until the first hybridization investigation carried out by Kölreuter [36]. With the discovery of the laws of heredity, in turn from 19th to 20th century, importance of hybridization in plant breeding became widely recognized. Therefore, understanding the environmental stress effects becomes vital for different cereal crop improvement programs which have depended mainly on the genetic variations present in the genome through conventional breeding. The development of elite salt-tolerant varieties of grain is considered the most cost-effective and environmentally safe method for further effective use of saline-alkali soils, it is important to maintain and preserve the genetic resources of agricultural crops.

Conventional breeding strategies are effective for improving tolerance to salinity, though success rate is low using traditional breeding methods. The mechanism of plant tolerance depends on physiological and genetic responses and involves screening genotypes that confer salt tolerance. Conventional breeding of crops play an important role in screening of genotypes for salinity tolerance and implicates crop improvement using selection, hybridization, polyploidy, and introgression proceedings. We can add simple trait with the help of backcross method to make an elite variety/cultivar and development of such an elite variety act like a backbone to perform conventional breeding processes. For instance, to develop a hybrid variety of the cross-pollinated crops numerous progressive methods are used like, recurrent selection, production of inbred lines, screening of the superior inbred lines and finally superior two inbred lines based on specific combining ability value [37]. Convention breeding has pre-requisite of the natural genetic variation existed for the desired trait. In cereals wheat is known for its diversity for the ion exchange mechanism of Na^+ and K^+/Na^+ ratio that exits in the form of landraces, progenitors [38], and other species having halophytic relationship in the family *triticeae* [39].

Targeted breeding for salt tolerance is done under coordinated wheat and barley program in India. However, there is a need to develop and exploit new sources of salt tolerant germplasm. Kharchia 65 is widely exploited genotype in India for the development of wheat varieties for salt tolerance and utilized as a donor parent for many wheat improvement programs globally [40]. However, it is not enough to consistently sustain the salt toxicity problem in all types of saline soils. The HD 2009 cultivar is considered as a susceptible parent to salinity that was released in 1975 for cultivation in North Western Plains of India grown under irrigated and timely sown conditions. In the recent past, wheat germplasm was screened for the identification of salt tolerant genotypes and only a few genotypes were identified in screening that imparts a significant level of tolerance against salt toxicity. Many wheat varieties were developed by employing identified tolerant genotypes and successfully cultivated throughout the world (**Table 1**). While, first salt tolerant variety of rice was basmati CSR 30 (Yamini) derived from the cross BR4-10/Pakistan Bas1, the donor BR4-10 from costal saline areas in state of Maharashtra, India.

Variety/line	Parents	Year	Developed by	Area of suitability	Reference
KH 65	Kharchia local/EG953	1970	Indian farmers through selection on sodicsaline Soil	Salinity soil of Durgapur, Rajasthan, India	[41]
Sakha 8	CNO67//SN64/KLRE/3/8156	1976	Agricultural Research Centre, Giza, Egypt	All saline soils	[42]
KRL1-4	WL711/Kharchia 65	1990	Central Soil Salinity Research Institute, Karnal, India	Saline/sodic soils of Northern India	[43]
KRL 19	PBW255/KRL 1-4	2000	Central Soil Salinity Research Institute, Karnal, India	Saline/sodic soils of Northern India	[43]
KRL 210	PBW65/2*PASTOR	2010	Central Soil Salinity Research Institute, Karnal, India	Sodic soil of Northern India	[44]
KRL 213	CNDO/R143//ENTE/MEXI_2/3/*Aegilops squarrosa* TAUS) /4/WEAVER/5/2*KAUZ	2010	Central Soil Salinity Research Institute, Karnal, India	Sodic soil of Northern India	[44]

Table 1.
Brief description about the development of salt tolerant wheat variety for using conventional breeding.

3.2 Molecular breeding approaches for salinity tolerance

To understand the inheritance and genomics, molecular breeding started by artificial crossing of parental lines. In modern breeding, several alleles are mixed together from different germplasm, and useful wild alleles are added to elite variety. In consort with the advances in high-throughput genotyping technology have enabled direct observation of alleles at loci. This permits understanding of allele's effects on particular phenotypes. Furthermore, it allows phenotypes prediction on the basis of genotypes with a genomic selection scheme. Finally, understanding the architecture of genome will help frame a precise breeding approach to deal with changes. On the other hand, polymorphic genetic factors numbers is too high in the genome to clarify their individual effects, particularly for quantitative traits. This confine power of genome-wide association studies to discover main quantitative trait loci. Genes and their molecular mechanisms for salt tolerance in cultivated crops should help breeders speed up genetic improvement of crop using marker-assisted selection (MAS) and genetic engineering [45]. Discovery of molecular markers is one of the most significant achievements of the biotechnology. They are extensively utilized to explore the DNA polymorphisms in the plant system and extremely useful for plant researchers and breeders to identify a particular trait linked with molecular markers at early stages of plants without doing phenotyping screening experiment [46]. To develop an elite line for salinity tolerance, identification of new QTLs is a key point. Salt tolerance is governed by more than one gene and is controlled by different QTL (quantitative trait loci) which are largely influenced by the environmental conditions. Thomson et al. [47] identified the "saltol" QTL that is involve into controls the Na^{+}/K^{+} ratio at the seedling stage in shoot using a conventional breeding approach in rice. According to Hasan et al. [48] one method of effective conventional breeding is marker-assisted backcrossing used to transfer alleles at target loci. Hence, the implementation of conventional breeding approaches shows limitations to handle the QTL character like salt tolerance in wheat crop improvement programs. Molecular markers are not influenced by the environmental conditions and they have been widely used for the QTL analyses in various mapping populations [49]. Molecular breeding approaches have different steps to go for the final level and track-down a single gene or QTL having linkage with the desired trait.

Some main QTLs and genes in crop plants associated with salinity tolerance are salt overly sensitive (SOS) at seedling, vegetative and reproductive stages which are hypersensitive to high external Na^{+}, Li^{+}, or K^{+} concentrations. These mutants are mutated at three loci *SOS1*, *SOS2*, and *SOS3* [50]. The *SOS1* encodes a plasma membrane $Na^{+/}H^{+}$ antiporter, *SOS2* activates the *SOS1* and encodes a serine/threonine protein kinase, and finally, *SOS3* gene encodes calcium-binding protein [51]. Single gene over expression could improve the tolerance to salinity of transgenic plants for example; *A. thaliana SOS1* and the vacuolar *AtNHX1* gene can considerably boost the salt tolerance of transgenic plants (**Table 2**). Moreover, at high salinity rice tolerance can improve using ABA-dependent regulatory pathways and high drought using *OsbZIP71* gene introducing the in transgenic plants. As per Su et al. [66], plant tolerance for salinity can improve by increasing the enzymes antioxidant activities and metabolism level of other mechanisms. The enzyme activity has also been confirmed in various transgenic plants by transferring bacterial genes. Plant resistance to oxidative stress improved through gene expression including; GAT, GR, SOD, and APX genes. The QTLs linked with grain yield were identified on chromosomes 1A, 1B, 2D, 3B, 4A, 4D, 6A, 6D, 7A and 7D tightly linked with different SSR marker in wheat for salt tolerance. The QTLs for TGW were identified on chromosomes 2B, 2D, 3B 7A and 7D. The QTL for tiller number (TN) and number of earhead (NE) identified on 4B, 4D and 1A, 2A, 2D, 4D, 5B 7A respectively tightly linked with SSR marker in wheat for salt tolerance (**Table 3**).

3.3 Genome-wide studies (GWAS) for salt tolerance

Genome wide analysis studies (GWAS) approach for tolerance to abiotic stress has popular in last couple of years [67, 83]. GWAS is frequently being used to find and describe genetic basis of agronomic traits, which are generally influenced by numerous small genes [84]. GWAS identify single nucleotide polymorphism (SNP) variations and functional effects that best way to develop elite variety for salinity [85]. GWAS used to selection for natural variety population for genotyping on the basis of phenotypic variation. To find out the association among genetic loci and phenotypic variations in natural populations genotype, linkage disequilibrium (LD) analysis uses and it provide an important alternative to linkage mapping. In different abiotic and biotic stress conditions to discover the targeted gene using GWAS approach has led to identify the polymorphisms and identify the genetic loci which are accountable for

Crop	Gene family	Locus name/ Gene	Function/ Cellular response	Linked marker to gene	Reference
Transporters					
Wheat (*Triticum aestivum* L.)	HKT family	Nax1(HKT7, TmHKT1;4)-2AL chromosome	Involved in remove of Na^+ from xylem in roots and leaf sheath	Gwm312	[52, 53]
		Nax2 (HKT8, TmHKT1;5)-5AL chromosome	removes Na^+ from the xylem in the roots and enhancing K^+ loading into the xylem	gwm291, gwm410, gpw2181	[54]
		Kna1 (HKT8, TaHKT1;5)-4DL chromosome	Removes Na^+ from the leaves. Controls the selectivity of Na^+ and K^+ transport from root to shoot and maintains high K^+/Na^+ ratio	Xwg199, Xabc305, Xbcd402, Xpsr567 and Xpsr375	[55, 56]
		HKT1	Na^+-K^+-symporter, Salt stress		[57]
Rice *(Oryza sativa)*	Na^+-K^+-symporter	HKT1	Salt stress		[58]
Maize *(Zea mays L.)*	Na^+-H^+-dependent K^+ transporter	ZmHKT1	Salt stress		[59]
Antioxidants					
Wheat	Catalase	CAT	Drought stress		[60]
Rice	Ascorbate peroxidase	APX	Drought, Salt and Cold		[61]
	Superoxide dismutase	SOD	Abiotic stress		[62]
Osmolytes					
Wheat	Proline	P5CS	Drought		[63]
Rice	Glycine betaine	BADH	Heavy metal stress		[64]
Maize	Glycine betaine	bet A	Abiotic stress		[65]

Table 2.
Brief description about the genes underlying salt tolerance in cereal crops.

Traits	Chromosome	Markers name	Markers type	Population	Reference
Na^+	2A	Jagger_c4026_328	SNP	Genotype	[67]
	2A	Wmc272/Barc349	SSR		[68]
	2B	cfd73.1	SSR	RIL	[69]
	2B	wPt-4647/wmc147	SSR		[68]
	3A	wsnp_Ex_rep_ c106152_90334299	SNP	Genotype	[67]
		wpt-666,438	DArt	RIL	[70]
	3B	gwm493	SSR	RIL	[69]
		wpt-8303	DArt	RIL	[71]
	4B	gwm368	SSR	RIL	[72]
	5A	gwm205	SSR	RIL	[71]
	5B	RAC875_c28831_558	SNP	Genotype	[67]
	5D	gwm174	SSR	RIL	[72]
	6B	wsnp_Ex_ c45713_51429315	SNP	Genotype	[67]
	7A	barc121	SSR	RIL	[69]
		gwm282	SSR	RIL	[71]
		wmc0017	SSR	DH	[73]
K^+	1D	RAC875_c14137_994	SNP	Genotype	[67]
		barc169	SSR	RIL	[69]
	2A	wpt-1142, wpt-4199	DArt	RILs	[71]
		wpt-4559, wpt-3378	DArt	RILs	[70]
		barc1155	SSR	RILs	[69]
		X1103701.44AG	SNP	DH	[73]
	2D	gwm132	SSR	RIL	[72]
		gwm261	SSR	RIL	[74]
	3B	wpt-0302, wpt-0895	DArt	RILs	[71]
		barc251	SSR	RIL	[69]
	3D	gwm191	SSR	RIL	[72]
	4B	barc193	SSR	RIL	[69]
	5A	IAAV8258	SNP	Genotype	[67]
		Vrn-A1	SNP	DH	[73]
		barc151	SSR	RIL	[69]
	5D	RAC875_rep_ c70595_321	SNP	Genotype	[67]
		gwm174	SSR	RIL	[72]
	6B	Kukri_c49331_77	SNP	Genotype	[67]
	7D	Excalibur_c13094_523	SNP	Genotype	[67]

Traits	Chromosome	Markers name	Markers type	Population	Reference
TN	4B	gwm6	SSR	Genotype	[68]
	4D	cfd84	SSR	RIL	[74]
NE	1A	gwm71, wmc59	SSR	DH	[75]
	2A	gwm71	SSR	DH	[75]
	2D	cfd53	SSR	DH	[75]
	4D	cfd84	SSR	RILs	[74]
	5B	gwm 499	SSR	RILs	[72]
	7A	gwm635	SSR	DH	[76]
DTH	2A	wmc177(Drought stress)	SSR	RIL	[77]
	2D	wmc112	SSR	RIL	[73]
	7D	XC29-P13(Drought stress)	SNP	RIL	[78]
DTA	7D	X7D-acc/cat-10(Drought stress)	SNP	RIL	[78]
	2D	wmc112	SSR	RIL	[74]
TGW	2B	gwm55	SSR	DH	[75]
	2D	wPt-8330, wPt-666,857	DArt	RILs	[71]
		cfd53	SSR	DH	[75]
		gwm 296 and wmc 601	SSR	DH	[53, 76]
	3B	gwm383	SSR	DH	[53, 76, 79]
		gwm 247	SSR	RIL	[80]
	7A	gwm282	SSR	DH	[76, 80]
	7D	barc 172	SSR	DH	[79, 81]
GY	1A	gwm 357	SSR	RIL	[80]
	1B	gwm11	SSR	DH	[81]
	1D	gwm642	SSR	RIL	[72]
	2D	gwm311	SSR	DH	[81]
		wmc41	SSR	RIL	[82]
		gwm261	SSR	RIL	[82]
		wPt-4413	DArt	RIL	[80]
		wmc601	SSR	RIL	[74]
	3B	gwm566,Gwm247	SSR	RILs	[72, 80]
		gwm 247	SSR	RIL	[80]
	3D	gwm314, gwm645	SSR	DH, RILs	[72, 81]
	4A	wpt-4620	DArt	RIL, Germplasm	[80]
	4D	Gwm194	SSR	DH	[81]
	6A	XP02m22-3, gwm169	RFLP, SSR	DH	[75]
	6D	gwm469	SSR	DH	[75]
	7A	gwm282	SSR	RIL	[79, 80, 82]
	7D	gwm437	SSR	RIL	[74]

Table 3.
Brief description about the identified QTLs linked to salt stress tolerance in wheat.

phenotypic variances [57, 86]. High salinity increased osmotic pressure into the soil and cause drought condition dropping, water absorption by the seed to the soil resulting, delayed seed germination [87]. Using different sequencing platform development of SNP marker is a modern technology for crop improvement. More than thousands SNPs are available has directed to the use of GWAS method in cereal crops to dissever traits. Association mapping based on candidate gene (CG) have targeted grain yield and yield related traits and physiological traits [88]. Edae et al. [89] identified three candidate (i.e. *DREB1A*, *ERA1* and *1-FEH*) genes in wheat with multiple agronomic and physiological traits using SNPs association and CG-association mapping. Moreover marker trait association (MTAs) for heat tolerance in wheat at seedling stage first reported by Maulana et al. [90] and found QTL on chromosomes 3B and 4B. Schmidt et al. [91] identified QTL for heat and drought tolerance in spring wheat using GWAS. Qaseem et al. [92] reported stable association on 5A and 7D chromosome for drought tolerance. Furthermore, significant association of yield traits for drought and heat tolerance was identified on 6A chromosome [91]. Qaseem et al. [92] find out three haplotypes on 1A, 3B and 6B chromosomes for salt tolerance index in 307 wheat accessions by affymetrix wheat 660 K SNP array. The QTLs for yield and other associated traits identified on 4A, 5A, 5B, 6B and 7A chromosome for salt tolerance [93].

4. Conclusion and future perspectives

In the wake of global warming and climate change, abiotic stresses like salinity is a major constrain to agriculture and allied sector which affects sustainable food and nutrition security. During last decades considerable efforts have been made to decipher mechanism of stress resistance and metabolic processes against various environmental factors affecting plant growth and yield potential (**Figure 3**). The degree of salt stress and plant growth stage are mainly responsible for foreseeing how plants defend and respond to prevailing stress conditions. In terms of drought, stomata close progressively along with a parallel reduction in water-use efficiency and net photosynthetic activity. Under drought conditions, lower stomatal conductivity and moisture use-efficiency leads to the impaired photosynthetic potential and overall growth in plants. In addition to several aspects, changes of plant pigments were found to be closely related to salinity tolerance in crops. In plants, self-defense mechanism operated at cellular level in leaf is triggered frequently to protect entire solar energy trapping system existing in the form of photosynthetic machinery from permanent damage. Removal of reactive-oxygen species through multiple enzymatic and non-enzymatic antioxidant defense pathways, cellular transportation and membrane stability, expression of underlying genes networks, and biosynthesis of array of defensins are key mechanisms of salinity tolerance.

Several plant genetic and genomic resources have been generated to address the abiotic stresses by different scientific groups under independent and coordinated research programs globally. With the use of conventional plant breeding approaches, multiple crop varieties, elite cultivars, promising lines and germplasm collections and gene bank accessions have been developed and preserved for cereal crop species. To support the traditional phenotypic selection-based breeding, advanced high-throughput phenotypic tools have also been devised and being utilized in phenotyping of various phenotypic traits manifested under stress conditions. Similarly, with advancement and availability of the next-generation sequencing technology at affordable cost, has paved the way to develop high-throughput genotyping platforms that used in whole-genome sequencing. The NGS-based genomic tools including SNP markers, genome maps, genome-wide QTLs and MTAs and gene networks underlying abiotic stress resistance have been deciphered and implemented in crop improvement programs to achieve higher genetic gains. Despite of technological advancements and resource mobility, usually a significant gap in terms of mutual cooperation and liaising needed in lab to land technology transfer have been witnessed at both institutional and scientific levels. Moreover, the potential of untapped alleles occurring in wild genetic stocks and genomic advancement are yet to be realized adequately in plant breeding.

Author details

Ram Baran Singh[1*] and Rajni Devi[2]

1 Department of Biotechnology, School of Applied Sciences (SoAS), REVA University, Bangalore, Karnataka, India

2 Division of Biochemistry, ICAR- IIWBR – Indian Institute of Wheat and Barley Research, Karnal, Haryana, India

*Address all correspondence to: rbsingh0611@gmail.com

References

[1] Fernie J, Pitkethly AS. Resources, Environment and Policy. First Published in 1985 by Harper & Row Ltd. Spitalfields, London: Reprinted in this edition by Paul Chapman Publication Ltd; 1985. pp. 306-308

[2] Wrigley CW. An overview of the family of cereal grains prominent in world agriculture. In: Encyclopedia of Food Grains. Vol. 1. Amsterdam, The Netherlands: Elsevier; 2016. pp. 73-85

[3] Srivastava RK, Singh RB, Srikanth B, Satyavathi CT, Yadav R, Gupta R. Genome-wide association studies (GWAS) and genomic selection (GS) in pearl millet: Advances and prospects. Frontiers in Genetics. 2020;**10**:1389

[4] Olugbire O, Olorunfemi S, Oke D. Global utilization of cereals: Sustainability and environmental issues. Agricultural Science. 2021;**20**(1):9-14

[5] Cuong DM, Kwon SJ, Nguyen BV, Chun SW, Kim JK, Park SU. Effect of salinity stress on phenylpropanoid genes expression and related gene expression in wheat sprout. Agronomy. 2020;**10**(3):390

[6] Szira F, Monostori I, Galiba G, Rakszegi M, Bálint AF. Micronutrient contents and nutritional values of commercial wheat flours and flours of field-grown wheat varieties—A survey in Hungary. Cereal Research Communications. 2014;**42**:293-302

[7] Cakmak I, Kirkby EA. Role of magnesium in carbon partitioning and alleviating photooxidative damage. Physiology Plant. 2008;**133**:692-704

[8] Rakascan N, Drazic G, Zivanovic L, Ikanovic J, Jovovic Z, Loncar M, et al. Effect of genotypes and locations on wheat yield components. The Journal of Agriculture and Forestry. 2019;**65**(1):233-242

[9] FAO. FAO Land and Plant Nutrition Management Service. 2022. Available from: http://www.fao.org/ag/agl/agII/spush

[10] Burgarella C, Cubry P, Kane NA, Varshney RK, Mariac C, Liu X, et al. A western Sahara Centre of domestication inferred from pearl millet genomes. Nature Ecology and Evolution. 2018;**2**(9):1377-1380

[11] FAO. The State of Food and Agriculture. 2021. Available from: https://www.fao.org/publications/sofa/sofa-2021/en/

[12] Munns RM, Tester. Mechanisms of salinity tolerance. Annual Review of Plant Biology. 2008;**59**:651-681

[13] Rao YR, Ansari MW, Sahoo RK, Wattal RK, Tuteja N, Kumar VR. Salicylic acid modulates *ACS*, *NHX1*, *sos1* and *HKT1*;2 expression to regulate ethylene overproduction and Na^+ ions toxicity that leads to improved physiological status and enhanced salinity stress tolerance in tomato plants cv. Pusa Ruby. Plant Signaling & Behavior. 2021;**16**(11):1950888

[14] Singh RB, Mahenderakar MD, Jugran AK, Singh RK, Srivastava RK. Assessing genetic diversity and population structure of sugarcane cultivars, progenitor species and genera using microsatellite (SSR) markers. Gene. 2020;**753**:144800

[15] Jeyasri R, Muthuramalingam P, Satish L, Pandian SK, Chen JT,

Ahmar S, et al. An overview of abiotic stress in cereal crops: Negative impacts, regulation, biotechnology and integrated omics. Plants. 2021;**10**(7):1472

[16] Arzani A, Ashraf M. Cultivated ancient wheat (*Triticum* spp.): A potential source of health-beneficial food products. Comprehensive Reviews in Food Science and Food Safety. 2017;**16**(3):477-488

[17] Kumar P, Sharma PK. Soil salinity and food security in India. Frontiers in Sustainable Food Systems. 2020;**4**:533781

[18] Sharma RC, Rao BRM, Saxena RK. Salt affected soils in India current assessment. In: Advances in Sodic Land Reclamation. Proceedings of the International Conference on Sustainable Management of Sodic Lands. Lucknow, India; 2004. pp. 1-26

[19] FAO. FAO Land and Plant Nutrition Management Service. 2008. Available from: http://www.fao.org/ag/agI/agII/spush

[20] FAO. World food situation. FAO Cereal Supply and Demand Brief. 2020. Available from: http://www.fao.org/worldfoodsituation/csdb/en/ [Accessed: August 15, 2020]

[21] Rengasamy P, Chittleborough D, Helyar K. Root zone constraints and plant-based solutions for dryland salinity. Plant and Soil. 2003;**257**:249-260

[22] Abbasi H, Jamil M, Haq A, Ali S, Ahmad R, Malik Z, et al. Salt stress manifestation on plants, mechanism of salt tolerance and potassium role in alleviating it: A review. Zemdirbyste-Agriculture. 2016;**103**(2):229-238

[23] Munns R, Schachtman DP. Plant responses to salinity: Significance in relation to time. International Crop Science. 2015;**2015**:741-745

[24] Rana V, Ram S, Nehra K, Sharma I. Expression of genes related to Na^+ exclusion and proline accumulation in tolerant and susceptible wheat genotypes under salt stress. Cereal Research Communications. 2016;**44**(3):404-413

[25] Roy SJ, Nagrao S, Tester M. Salt resistant crop plants. Current Opinion in Biotechnology. 2014;**26**:115-124

[26] Khan A, Khan AL, Muneer S, Kim YH, Al-Rawahi A, Al-Harrasi A. Silicon and salinity: Crosstalk in crop-mediated stress tolerance mechanisms. Frontiers in Plant Science. 2019;**10**:1429

[27] Schachtman DP, Schroeder JI. Structure and transport mechanism of a high-affinity potassium uptake transporter from higher plants. Nature. 1994;**370**:655-658

[28] El-Hendawy SE, Hassan WM, Al-Suhaibani NA, Refay Y, Abdella KA. Comparative performance of multivariable agro-physiological parameters for detecting salt tolerance of wheat cultivars under simulated saline field growing conditions. Frontiers in Plant Science. 2017;**8**(435):1-15

[29] Mbarki S, Sytar O, Zivcak M, Abdelly C, Cerda A, Brestic M. Anthocyanins of coloured wheat genotypes in specific response to salstress. Molecules. 2018;**23**:1518

[30] Munns R, Wallace PA, Teakle NL, Colmer TD. Measuring soluble ion concentrations (Na^+, K^+, Cl^-) in salt-treated plants. In: Plant Stress Tolerance: Methods and Protocols. Vol. 23. Basel, Switzerland: Springer Nature; 2010. pp. 371-382

[31] Kumar D, Al Hassan M, Naranjo MA, Agrawal V, Boscaiu M, Vicente O. Effects of salinity and drought on growth, ionic relations, compatible solutes and activation of antioxidant systems in oleander (*Nerium oleander* L.). PLoS One. 2017;**12**(9):e0185017

[32] Zivcak M, Brestic M, Olsovska K, Slamka P. Performance index as a sensitive indicator of water stress in *Triticum aestivum*. Plant Soil Environment. 2008;**54**:133-139

[33] Ullah H, Gul B, Khan H, Zeb U. Effect of salt stress on proximate composition of duckweed (*Lemna minor* L.). Heliyon. 2021;**7**(6):e07399

[34] Flowers TJ, Colmer TD. Salinity tolerance in halophytes. New Phytologist. 2008;**2008**:945-963

[35] Ashraf M, Foolad MA. Improving plant abiotic-stress resistance by exogenous application of osmo-protectants glycine betaine and proline. Environmental and Experimental Botany. 2007;**59**:206-216

[36] Kölreuter JG. Vorläufige Nachricht von einigen das Geschlecht der pflanzen betreffenden Versuchen und Beobachtungen, nebst Fortsetzungen 1, 2 und 3 (1761-1766). Leipzig: Wilhelm Engelmann; 1893

[37] Janaki Ramayya P, Vinukonda VP, Singh UM, Alam S, Venkateshwarlu C, Vipparla AK, et al. Marker-assisted forward and backcross breeding for improvement of elite Indian rice variety Naveen for multiple biotic and abiotic stress tolerance. PLoS One. 2021;**16**(9):e0256721

[38] Gorham J. Genetics and physiology of enhanced K/Na discrimination. In: Randall P, editor. Genetic Aspects of Plant Mineral Nutrition. Dordrecht, The Netherlands: Kluwer Academic Publishers; 1993. pp. 151-159

[39] Garthwaite AJ, von Bothmer R, Colmer TD. Salt tolerance in wild *Hordeum* species is associated with restricted entry of Na^+ and Cl^- into the shoots. Journal of Experimental Botany. 2005;**56**(419):2365-2378

[40] Shahbaz M, Ashraf M. Improving salinity tolerance in cereals. Critical Reviews in Plant Sciences. 2013;**32**(4):237-249

[41] Hu J, Wang X, Zhang G, Jiang P, Chen W, Hao Y, et al. QTL mapping for yield-related traits in wheat based on four RIL populations. Theoretical and Applied Genetics. 2020;**133**(3):917-933

[42] Rana RS. Evaluation and utilization of traditionally grow cereals cultivars on salt affected area in India. Indian Journal of Genetics and Plant Breeding. 1986;**46**:121-135

[43] Munns R, Richard AJ, Lauchli A. Approaches to increasing the salt tolerance of wheat and other cereals. Journal of Experimental Botany. 2006;**57**:1025-1043

[44] Hollington PA. Technological breakthroughs in screening/breeding wheat varieties for salt tolerance. In: Gupta SK, Sharma SK, Tyagi NK, editors. Proceedings of the National Conference 'Salinity Management in Agriculture'. Karnal India: Central Soil Salinity Research Institute; 2000. pp. 273-289

[45] Hanin M, Ebel C, Ngom M, Laplaze L, Masmoudi K. New insights on plant salt tolerance mechanisms and their potential use for breeding. Frontiers in Plant Science. 2016;**7**:1787

[46] Soto-Cerda BJ, Cloutier S. Association mapping in plant genomes.

In: Caliskan M, editor. Genetic Diversity in Plants. London, United Kingdom: IntechOpen Limited; 2012. pp. 29-54. ISBN: 978-953-51-0185-7

[47] Thomson MJ, de Ocampo M, Egdane J, Rahman MA, Sajise AG, Adorada DL, et al. Characterizing the Saltol quantitative trait locus for salinity tolerance in rice. Rice. 2010;**3**(2):148-160

[48] Hasan M, Abdullah HM, Hasibuzzaman ASM, Ali MA. Date palm genetic resources for breeding. In: Cash Crops: Genetic Diversity, Erosion, Conservation and Utilization. Basel, Switzerland: Springer Nature; 2022. pp. 479-503

[49] Kumar S, Röder MS, Tripathi SB, Kumar S, Chand R, Joshi AK, et al. Mendelization and fine mapping of a bread wheat spot blotch disease resistance QTL. Molecular Breeding. 2015;**35**(11):1-10

[50] Martínez-Atienza J, Jiang X, Garciadeblas B, Mendoza I, Zhu JK, Pardo JM, et al. Conservation of the salt overly sensitive pathway in rice. Plant Physiology. 2007;**143**(2):1001-1012

[51] Mahajan S, Pandey GK, Tuteja N. Calcium-and salt-stress signaling in plants: Shedding light on SOS pathway. Archives of Biochemistry and Biophysics. 2008;**471**(2):146-158

[52] Lindsay MP, Lagudah ES, Hare RA, Munns R. A locus for sodium exclusion (Nax1), a trait for salt tolerance, mapped in durum wheat. Functional Plant Biology. 2004;**31**:1105-1114

[53] Huang S, Spielmeyer W, Lagudah ES, James RA, Platten JD, Dennis ES, et al. A sodium transporter (HKT7) is a candidate for Nax1, a gene for salt tolerance in durum wheat. Plant Physiology. 2006;**142**:1718-1727

[54] Byrt CS, Platten JD, Spielmeyer W, James RA, Lagudah ES, Dennis ES, et al. HKT 1;5-like cation transporters linked to Na^+ exclusion loci in wheat, Nax2 and KNa1. Plant Physiology. 2007;**143**:1918-1928

[55] Gorham J, Jones RG, Bristol A. Partial characterization of the trait for enhanced K^+-Na^+ discrimination in the D genome of wheat. Planta. 1990;**180**:590-597

[56] Dubcovsky J, Maria SG, Epstein E, Luo MC, Dvořák J. Mapping of the K+/Na+ discrimination locus Kna1 in wheat. Theoretical and Applied Genetics. 1996;**2**:448-454

[57] Xu Y, Yang T, Zhou Y, Yin S, Li P, Liu J, et al. Genome-wide association mapping of starch pasting properties in maize using single-locus and multi-locus models. Frontiers in Plant Science. 2018;**9**:1311

[58] Basu S, Roychoudhury A. Expression profiling of abiotic stress-inducible genes in response to multiple stresses in rice (*Oryza sativa* L.) varieties with contrasting level of stress tolerance. BioMed Research International. 2014;**2014**(706890):1-12

[59] Jiang Z, Song G, Shan X, Wei Z, Liu Y, Jiang C, et al. Association analysis and identification of *ZmHKT1*; 5 variation with salt-stress tolerance. Frontiers in Plant Science. 2018;**12**(9):1485

[60] Luna CM, Pastori GM, Driscoll S, Groten K, Bernard S, Foyer CH. Drought controls on H_2O_2 accumulation, catalase (CAT) activity and CAT gene expression in wheat. Journal of Experimental Botany. 2005;**56**(411):417-423

[61] Zhang Z, Zhang Q, Wu J, Zheng X, Zheng S, Sun X, et al. Gene knockout study reveals that cytosolic ascorbate

peroxidase 2 (OsAPX2) plays a critical role in growth and reproduction in rice under drought, salt and cold stresses. PLoS One. 2013;**8**(2):e57472

[62] Shiraya T, Mori T, Maruyama T, Sasaki M, Takamatsu T, Oikawa K, et al. Golgi/plastid-type manganese superoxide dismutase involved in heat-stress tolerance during grain filling of rice. Plant Biotechnology Journal. 2021;**13**(9):1251-1263

[63] Maghsoudi K, Emam Y, Niazi A, Pessarakli M, Arvin MJ. P5CS expression level and proline accumulation in the sensitive and tolerant wheat cultivars under control and drought stress conditions in the presence/absence of silicon and salicylic acid. Journal of Plant Interactions. 2018;**13**(1):461-471

[64] Paul S, Banerjee A, Roychoudhury A. Role of polyamines in mediating antioxidant defense and epigenetic regulation in plants exposed to heavy metal toxicity. In: Plants under Metal and Metalloid Stress. Singapore: Springer; 2018. pp. 229-247

[65] Quan R, Shang M, Zhang H, Zhao Y, Zhang J. Engineering of enhanced glycine betaine synthesis improves drought tolerance in maize. Plant Biotechnology Journal. 2004;**2**(6):477-486

[66] Su Y, Guo A, Huang Y, Wang Y, Hua J. GhCIPK6a increases salt tolerance in transgenic upland cotton by involving in ROS scavenging and MAPK signaling pathways. BMC Plant Biology. 2020;**20**(1):1-19

[67] Oyiga BC, Sharma RC, Baum M, Ogbonnaya FC, Léon J, Ballvora A. Allelic variations and differential expressions detected at quantitative trait loci for salt stress tolerance in wheat. Plant, Cell & Environment. 2018;**41**(5):919-935

[68] Genc Y, Oldach K, Verbyla A, Lott G, Hassan M, Tester M, et al. Sodium exclusion QTL associated with improved seedling growth in bread wheat under salinity stress. Theoretical and Applied Genetics. 2010;**121**:877-894

[69] Xu YF, An DG, Liu DC, Zhang AM, Xu HX, Li B. Mapping QTLs with epistatic effects and QTL× treatment interactions for salt tolerance at seedling stage of wheat. Euphytica. 2012l;**186**(1):233-245

[70] XuY ZY, Hong S, Xia Z, Cui D, Guo J, Xu H, et al. Functional characterization of a wheat NHX antiporter GeneTaNHX2 that encodes a K^+/H^+ exchanger. PLoS One. 2013;**8**(11):78-98

[71] Masoudi B, Mardi M, Hervan EM, Bihamta MR, Naghavi MR, Nakhoda B, et al. QTL mapping of salt tolerance traits with different effects at the seedling stage of bread wheat. Plant Molecular Biology Reporter. 2015;**33**(6):1790-1803

[72] Amin AY, Diab AA. QTL mapping of wheat (*Triticum aestivum* L.) in response to salt stress. International Journal of Biotechnology Research. 2013;**3**:47-60

[73] Asif MA, Schilling RK, Tilbrook J, et al. Mapping of novel salt tolerance QTL in an Excalibur × Kukri doubled haploid wheat population. Theoretical and Applied Genetics. 2018;**131**:2179-2196

[74] Devi R, Ram S, Rana V, Malik VK, Pande V, Singh GP. QTL mapping for salt tolerance associated traits in wheat (*Triticum aestivum* L.). Euphytica. 2019;**215**(12):1-23

[75] Heidari B, Sayed-Tabatabaei BE, Saeidi G, Kearsey M, Suenaga K. Mapping QTL for grain yield, yield components, and spike features in a

doubled haploid population of bread wheat. Genome. 2011;**54**(6):517-527

[76] Huang XQ, Kempf H, Ganal MW, Röder MS. Advanced backcross QTL analysis in progenies derived from a cross between a German elite winter wheat variety and a synthetic wheat (*Triticum aestivum* L.). Theoretical and Applied Genetics. 2004;**109**:933-943

[77] Maccaferri M, Mantovani P, Tuberosa R, DeAmbrogio E, Giuliani S, Demontis A, et al. A major QTL for durable leaf rust resistance widely exploited in durum wheat breeding programs maps on the distal region of chromosome arm 7BL. Theoretical and Applied Genetics. 2008;**117**(8):1225-1240

[78] Lopes MS, Silva FF, Harlizius B, Duijvesteijn N, Lopes PS, Guimarães SE, et al. Improved estimation of inbreeding and kinship in pigs using optimized SNP panels. BMC Genetics. 2013;**14**(1):1-1

[79] Gupta PK, Balyan HS, Gahlaut V. QTL analysis for drought tolerance in wheat: Present status and future possibilities. Agronomy. 2017;**7**(1):5

[80] Azadi A, Mardi M, Hervan EM, Mohammadi SA, Moradi F, Tabatabaee MT, et al. QTL mapping of yield and yield components under normal and salt-stress conditions in bread wheat (*Triticum aestivum* L.). Plant Molecular Biology Reporter. 2015;**33**(1):102-120

[81] Kuchel H, Williams KJ, Langridge P, Eagles HA, Jefferies SP. Genetic dissection of grain yield in bread wheat. I. QTL analysis. Theoretical and Applied Genetics. 2007;**115**(8):1029-1041

[82] Kumar N, Kulwal PL, Balyan HS, Gupta PK. QTL mapping for yield and yield contributing traits in two mapping populations of bread wheat. Molecular Breeding. 2007;**19**(2):163-177

[83] Sehgal D, Rosyara U, Mondal S, Singh R, Poland J, Dreisigacker S. Incorporating genome-wide association mapping results into genomic prediction models for grain yield and yield stability in CIMMYT spring bread wheat. Frontiers in Plant Science. 2020;**11**:197

[84] Yu J, Zhao W, Tong W, He Q, Yoon MY, Li FP, et al. A genome-wide association study reveals candidate genes related to salt tolerance in rice (*Oryza sativa*) at the germination stage. International Journal of Molecular Sciences. 2018;**19**(10):3145

[85] Afzal M, Hindawi SE, Alghamdi SS, Migdadi HH, Khan MA, Hasnain MU, et al. Potential breeding strategies for improving salt tolerance in crop plants. Journal of Plant Growth Regulation. 2023;**42**(6):3365-3387

[86] Afzal M, Alghamdi SS, Nawaz H, Migdadi HH, Altaf M, El-Harty E, et al. Genome-wide identification and expression analysis of CC-NB-ARC-LRR (NB-ARC) disease-resistant family members from soybean (Glycine max L.) reveal their response to biotic stress. Journal of King Saud University-Science. 2022;**34**(2):101758

[87] Sayar R, Bchini H, Mosbahi M, Ezzine M. Effects of salt and drought stresses on germination, emergence and seedling growth of durum wheat (*Triticum durum* Desf.). Journal of Agriculture Research. 2010;**5**(15):2008-2016

[88] Afzal F, Li H, Gul A, Subhani A, Ali A, Mujeeb-Kazi A, et al. Genome-wide analyses reveal footprints of divergent selection and drought adaptive traits in synthetic-derived

wheats. G3: Genes, Genomes, Genetics. 2019;**9**(6):1957-1973

[89] Edae EA, Byrne PF, Manmathan H, Haley SD, Moragues M, Lopes MS. Association mapping and nucleotide sequence variation in five drought tolerance candidate genes in spring wheat. Plant Genome. 2013;**6**:13

[90] Maulana F, Ayalew H, Anderson JD, Kumssa TT, Huang W, Ma XF. Genome-wide association mapping of seedling heat tolerance in winter wheat. Frontiers in Plant Science. 2018;**9**:1272

[91] Schmidt J, Tricker PJ, Eckermann P, Kalambettu P, Garcia M, Fleury D. Novel alleles for combined drought and heat stress tolerance in wheat. Frontiers in Plant Science. 2020;**10**:1800

[92] Qaseem MF, Qureshi R, Muqaddasi QH, Shaheen H, Kousar R, Röder MS. Genome-wide association mapping in bread wheat subjected to independent and combined high temperature and drought stress. PLoS One. 2018;**13**(6):e0199121

[93] Yu S, Wu J, Wang M, Shi W, Xia G, Jia J, et al. Haplotype variations in QTL for salt tolerance in Chinese wheat accessions identified by marker-based and pedigree-based kinship analyses. The Crop Journal. 2020;**8**(6):1011-1024